Digital Value Migration in Media, ICT and Cultural Industries

Societies today are in a period of dynamic change, highly fluid and contested in moving from traditional to liberal and from local to global, as well as ranging from highly developed to emerging market economies. Alongside and facilitating this is a rapidly and exponentially changing digital media industry, including new technologies, multi-platform distributions and advertising models.

This monograph highlights, identifies, evaluates and provides rich insight into the complex nature and meaning of different digital value migration in media corporations and ICT companies. It illustrates how such values affect both the internal and the external environments of media companies and industries, as well as prosumers' consumption. Including chapters from expert scholars and industry practitioners representing cutting-edge research in the U.S. and Europe in the fields of digital convergence, broadband, media and information communication technology (ICT) business and technology, the book helps academics, researchers, media policymakers and corporate executives better understand today's undulating media and ICT markets. Specifically, it illuminates where they have come from, what is at stake and what forces drive and constrain them in global hypercompetitive markets. Ultimately, it aims relatedly to facilitate high academic, business and professional standards.

This text will be of key interest to scholars, students and business and industry practitioners in digital media, media management, international business, media economics and media policy and, more broadly, to those in the cultural industries, strategic management, business studies and marketing.

Zvezdan Vukanovic is Associate Professor at Abu Dhabi University, United Arab Emirates.

Mike Friedrichsen is President of the Berlin University of Digital Sciences and Full Professor at the Stuttgart Media University, Germany.

Milivoje Pavlovic is Full Professor and Vice-Rector at Megatrend University in Belgrade, Serbia.

World Politics and Dialogues of Civilizations Series
Series Editor: Raffaele Marchetti
LUISS Guido Carli University, Italy

This new series aims to explore alternative models of the social, political and economic developments at regional and world levels in order to advance theoretical understanding, promote political debate and provide policy-oriented advice. It focuses on six macro areas: policies, institutions and shared prosperity; infrastructure for global inclusive development; the economy beyond the failure of conventional models; East and West, North and South; civilisations against the threat of social barbarism; and life space for humanity.

The series is linked to the research carried out by the Dialogue of Civilizations (DoC) Research Institute (Berlin) but is open to external contribution.

Debating Migration to Europe
Welfare vs Identity
Raffaele Marchetti

Liberal World Order and Its Critics
Civilisational States and Cultural Commonwealths
Adrian Pabst

Digital Value Migration in Media, ICT and Cultural Industries
From Business and Economic Models/Strategies to Networked Ecosystems
Edited by Zvezdan Vukanovic, Mike Friedrichsen and Milivoje Pavlovic

For more information on this series, please visit: https://www.routledge.com/World-Politics-and-Dialogues-of-Civilizations/book-series/DOC

Digital Value Migration in Media, ICT and Cultural Industries

From Business and Economic Models/ Strategies to Networked Ecosystems

Edited by
Zvezdan Vukanovic, Mike Friedrichsen and Milivoje Pavlovic

Routledge
Taylor & Francis Group
LONDON AND NEW YORK

D C RESEARCH INSTITUTE

First published 2019
by Routledge
2 Park Square, Milton Park, Abingdon, Oxon OX14 4RN

and by Routledge
52 Vanderbilt Avenue, New York, NY 10017

First issued in paperback 2020

*Routledge is an imprint of the Taylor & Francis Group,
an informa business*

British Library Cataloguing in Publication Data
A catalogue record for this book is available from the British Library

Library of Congress Cataloging in Publication Data
A catalog record has been requested for this book

ISBN 13: 978-0-367-66135-9 (pbk)
ISBN 13: 978-1-138-37012-8 (hbk)

Typeset in Times NR MT Pro
by Cenveo® Publisher Services

Contents

List of figures

List of tables

List of contributors

José María Álvarez-Monzoncillo Professor of Audiovisual Communications and Deputy Rector for International Affairs, Rey Juan Carlos University, Madrid. Email contact: josemaria.alvarez@urjc.es

John Carey Program Director, MS in Media Management, Fordham University, New York. Email contact: johncarey@fordham.edu

Guillermo de Haro-Rodríguez Associate Professor at EAE Business School, Madrid. Email contact: guillermo.deharo@gmail.com

Paulo Faustino Chairman of the International Media Management Academic Association (IMMAA). Email contact: faustino.paulo@gmail.com

Mike Friedrichsen Professor, Berlin University of Digital Sciences. Email contact: friedrichsen@berlin-university.digital

Javier López-Villanueva Associate Professor of Media Economics, Rey Juan Carlos University, Madrid. Email contact: javier.lopez@urjc.es

Vagia Mochla Researcher, School of Journalism and Mass Communications, Aristotle University of Thessaloniki. Email contact: vagiamochla@jour.auth.gr

Milica Nestorović Assistant Professor, Megatrend University, Belgrade. Email contact: mnestorovic@megatrend.edu.rs

Milivoje Pavlovic Professor, Megatrend University, Belgrade. Email contact: mpavlovic@megatrend.edu.rs

Tatjana Dragičević Radičević Associate Professor, Megatrend University, Belgrade. Email contact: tanjadragrad@hotmail.com

Miriam Rodríguez-Pallares Assistant Professor, Universidad Internacional de la Rioja, Madrid. Email contact: miriam.rodriguez@unir.net

George Tsourvakas Associate Professor, Aristotle University of Thessaloniki, Greece. Email contact: gtsourv@jour.auth.gr

Zvezdan Vukanovic Associate Professor, Abu Dhabi University. Email contact: zvezdan.vukanovic@adu.ac.ae

Foreword

The theme of *Digital Value Migration in Media, ICT and Cultural Industries* is appropriately chosen. Today, our societies are in a period of dynamic change, moving from traditional to liberal and from local to global, as well as ranging from highly developed to emerging market economies. Moreover, the information revolution influences our lifestyles, learning, work roles, social, economic infrastructure and even our opinions, attitudes and moral judgements.

Today, we are under the spell of information as it has become the key to influence and power. The originally developed demand for automatic sharing of knowledge and information between university scholars and research institute scientists around the world has revolutionized communications worldwide. Direct applications in the form of social media provide opportunities to create, share and exchange information between virtual communities, bringing about substantial changes in communications among individuals, organizations and communities.

With fast technological changes, this development has both blessing and drawback. Although, on the one side, there are numerous benefits for communication, entertainment and education via increasing the mobilization, transparency and commercial applications, we also see impediments in terms of control, invasion of privacy and a decrease of face-to-face interactions. As a result, the future is becoming more challenging.

In this dynamic technological, post-industrial era, foreseeing the future has always been a challenge for humankind not only in terms of progressive learning about contemporary media and communication industries, technologies and business models but also in regard to influencing the phenomena's positive outcome. This universal theme imposes a simple question: today, in this high civilization stage, is humankind able to choose and decide on its future, and, furthermore, is humankind capable of evading an unwanted future?

In sum, I wish to stress that this monograph will realise the goal of establishing a new global, coopetitive and networked media business culture together with effective ways of communicating it to people and various social and corporate digital media and ICT stakeholders.

Ljubisa Rakic
Member of the European Academy of Sciences and Arts (EASA)

Editorial note

This monograph attempts to establish and highlight a wide spectrum of the research agenda in *New Digital Value Migration in Media, ICT and Cultural Industries: From Business and Economic Models/Strategies to Networked Ecosystems.*

The book is trying to be eminently practical in its explanation of the evolutionary juxtaposition of media business and technological paradigms of the changing digital media scene from an interdisciplinary perspective, providing a number of major trends, determinants and activities/insights from business, economic, managerial, marketing, organizational, technological, entrepreneurial, intercultural and sociocultural approaches.

Importantly, this volume is the result of the editors' selection of 7 out of 55 papers presented at Megatrend University's international conference on Information Revolution, New Media and Social Changes in the World, which took the place in Belgrade on 25 November 2016. Accordingly, these selected papers were the most suitable to explore the characteristics, paradigms, dynamics, models, processes, challenges and implications of the development of ICT and digital media business convergence, models and networks in the 21st century.

Drawing upon contributions from academia and industry that represent current research in the U.S. and Europe, the editors have commissioned academic and business research papers from reputable, established and leading state-of-the-art international academic experts, scholars and industry practitioners in the fields of digital convergence, broadband, media and ICT business and technology to attain high academic, business and professional standards.

Despite the additional changes in ownership and corporate strategy that no doubt will take place in the next few years, we think the book will help academics, researchers and corporate executives better understand today's undulating media and ICT markets in terms of where they have come from, what is at stake and what forces drive and constrain them.

While it is perhaps unusual to cover so many disciplines when discussing any one topic, the editors believe that the broad impact and implications of digital media and ICT demand this approach. The ICT and media business is not, after all, just one industry. It is not only a communications medium but also an economic, social, scientific and educational vehicle. One monograph can only begin to explore this evolving topic.

In summary, this text is a primer on the issue of digital value migration in media, ICT and cultural industries in global hypercompetitive markets. Moreover, the authors hope that this book will stimulate a lively debate about this topic in Europe, North America, and beyond.

Zvezdan Vukanovic
Mike Friedrichsen
Milivoje Pavlovic

1   Economic and social patterns in the adoption of new media

John Carey

In the 21st century, we have experienced a rapid deployment of new media technologies. There have been more changes in the first 15 years of this century than in the previous 80 years. During the 20th century, there was generally one new technology introduced each decade: AM radio in the 1920s, FM radio in the 1930s, the LP in the 1940s, black-and-white TV in the 1950s, colour TV in the 1960s, VCRs in the 1970s, PCs and cell phones in the 1980s and the Internet and DVRs in the 1990s. Since the beginning of the 21st century, we have seen the introduction of HDTVs, smart TVs, 4K TVs, 3D TVs, broadband, Wi-Fi, tablets, smartphones, wearable technologies such as smart watches, the Internet of things (e.g., thermostats and other everyday devices connected to the Web) and voice recognition technologies, among others. With these recent technological changes has come a deluge of new services, including on-demand video, tweets, user-generated video, interactive video, immersive video, customised advertising, streaming music and apps for just about everything (there are over 2 million apps in the Apple App Store).

The methodology for this chapter utilises economic, historical and sociological methods. It draws from economic and social patterns associated with the modern era of rapid media deployment, both causes and effects, and compares these with some patterns in the 20th century. The economic analysis includes pricing, economic effects of rapid deployment, failures and audience measurement that affects advertising revenue. The sociological analysis includes the growing role of women and younger people in early adoption, mobility, baby boomers and technology, privacy and negativity in social media.

The analysis draws from historical data in both North America (Sterling and Haight, 1978) and Europe (Williams, 1978), as well as from more recent data about media usage patterns in North America (Pew, 2018) and Europe (Oxford Internet Institute, 2018). It also draws from classic models of media adoption (Rogers, 1995) and more recent models (Dutton, 2013).

1.1 Critical mass

The rapid pace of change and accompanying competition would seemingly make it difficult for many technologies and services to reach critical mass—a level of adoption where enough people have the technology that further adoption becomes self-sustaining (Rogers, 1995, 313). However, this does not appear to be the case; many (e.g., broadband, smartphones and social media) have reached critical mass. Perhaps technology use begets even more technology use. However, in the author's research (Carey and Elton, 2010) over the past 25 years, a few other factors are at work. First, people are more tech savvy than 20 years ago and can adopt new technology with greater ease. Second, people have developed many ways to 'filter' the mass of content and information that is available to them and get just what they want versus feeling overwhelmed with information, as were many in the 20th century. For example, apps and search engines like Google serve as important filters for people to control the information that reaches them (Herrman, 2016).

From a social perspective, there are many more women in the workforce, which creates a need for them to adopt many new technologies and media services that are necessary in the workplace. The workforce generally is more educated and better able to use advanced technologies. There are also more who work at home and thus a related need to adopt technologies that will support these activities. Also, many more people are in service jobs that require media technologies versus many more who were in agricultural jobs in the first quarter of the 20th century and manufacturing jobs in the second and third quarters of the century. And people are acquiring media technologies at a younger age. Many 8- to 10-year-olds have smartphones, and some children as young as 3 or 4 are using tablets to play games and watch videos (Pew, 2015).

Focusing on media use, there have been many changes that have helped media technologies to reach critical mass. One factor is that people are spending more time with a wide range of media. For example, the average American spends 12 hours a day exposed to media. However, this occurs within a 9-hour time frame (eMarketer, 2016). The extra hours of exposure are made possible by multitasking—e.g., using a smartphone while watching TV. The presence of media in so many public locations also makes it possible to use more technology. Some of this is TVs and information kiosks in public locations, but more is based on the mobile technologies that people carry with them all the time—e.g., smartphones, tablets and laptops. Even sleep patterns are affecting media adoption and use. Most people in the first quarter of the 21st century are sleeping less than people in the first quarter of the 20th century and therefore have more time to use technology (Brody, 2013).

1.2 Early adopters and price

A cartoon in *The New Yorker* in the 1970s nicely captured the characteristics of early adopters of new media at that time. In it, a middle-aged man is in an electronics store and says to the clerk, 'All my gadgets are old. I'd like some new gadgets.' No mention is made of price. Indeed, this was typical of many early adopters: they were male and middle-aged, had a high disposable income and liked technology for technology's sake.

In the 21st century, there have been important changes. The middle-aged, well-off gadget guy is still part of the mix of early adopters, but many more younger people and women are early adopters of new media. Further, many with less disposable income (or generous parents) are adopting new technology even at higher prices because technology is so core to their lives. One example is the iPhone. More younger people 18-29 own iPhones than all adults of any age taken together, and the number of women owning iPhones is roughly equal to that of males (Pew, 2015). While there are some 'gadget people' among this younger group and women, more use the technology because it has functional purposes (e.g., to find health information, the location of a restaurant, the weather) and because it is central to their lifestyles of anytime, anywhere access to media services (Katz, 2006, 3-14).

There is another group that is under-recognised as early adopters: baby boomers. While they are not as likely to be in the first wave of adopters compared to younger people and middle-aged gadget guys, they are present in very large numbers during the second wave (Pew, 2015). And, as a group, they have a great deal of disposable income. It is surprising that marketers of new media technologies do not seem to recognise this. How often do we see a commercial for the latest smartphone or tablet and the people in the commercial enjoying the device are in their 60s?

In general, the pattern of introducing new media devices at a high price and then dropping the price over time to achieve a mass audience has continued into the 21st century. HDTVs, smart TVs and 4K TVs followed this pattern. However, there are some exceptions, notably with Apple products. Apple has not dropped the price of its iPhone or iPad. Apple does sometimes offer a cheaper, stripped-down version of a product, as in the case of the iPad (Pro versus Mini models), but it has maintained the high price of its premier products.

Few people pay the full retail price for iPhones. Generally, the price is bundled with a mobile phone plan. In these cases, the mobile phone provider is subsidizing the price of the iPhone, with the cost to the provider returned from the monthly usage plan. Subsidies have also been built into the cost of some video game consoles in the expectation that the supplier will get the cost back through the sale of video game software.

Unlike media technologies, the cost of media services has generally not declined over time. For example, the cost of cable TV service, satellite TV service and satellite radio did not decline. The key variable distinguishing the two groups is content. Where there is no content or the service provider does not have to pay for content (as when users create it), it has been possible to bring down the cost of the service. When the service provides and pays for content, the costs of talent and production increase over time. Many other factors such as regulations and competition can affect the price of a service.

1.3 Replacement cycles

The adoption of one technology or service may be linked to the purchase of another. For example, sometimes people purchase a new technology because they want a service or experience they do not have. However, in other cases, they need to replace a technology they currently have that is out of date. In purchasing a replacement, there is an opportunity to upgrade and get the latest components and features. If the average replacement cycle is long (as in the case of TV sets), there is less opportunity to introduce new components and features. However, if the replacement cycle is short (as in the case of smartphones), new components and features can be introduced at a more rapid rate. This is a conservative model of adoption in which new technologies piggyback on replacement cycles. This pattern of adoption is sometimes called a Trojan Horse strategy. That is, if one technology can make its way into homes, it can open up the gates for others to follow, building upon the first technology. This is one reason that smartphone technology has advanced so quickly.

1.4 Failures, false starts and fads

There has been no shortage of failures, false starts and fads in the digital era. Failures and false starts can be difficult to distinguish in a first iteration of the technology or service. In the 20th century, laser videodiscs were a failure in the 1980s. However, a decade later the same basic technology was reborn, only now miniaturised and with much greater storage, as the successful DVD. So the laser videodisc was in fact a false start. In the 21st century, 3D TV has clearly failed. This has been due to a lack of content, the reluctance of people to wear 3D glasses while watching TV and nausea experienced by some from the 3D experience. However, there are a few follow-on technologies that could turn the original 3D TV into a false start. Virtual reality (VR) systems have been launched recently, and they can provide a 3D video experience. 3D TV without glasses is due in a few years (it is available now on some portable video game devices), and holographic TV is less than 10 years away. So the jury is out on whether 3D is a failure or a false start.

New media fads have been plentiful in this century, as they were in the past (remember the boom box in the 1990s?). Ring tones are generated when software downloaded to a mobile phone plays a song chosen by the owner to indicate an incoming phone call. They experienced a surge of interest for three or four years and then faded quickly. How often these days do we hear Elvis or Beyoncé singing from a person's pocket or purse to signal an incoming call? Most people now keep their phone on vibrate. The novelty of ring tones has also worn off because of an annoyance factor—e.g., when they went off during a meeting or important conversation. This created a social stigma for ring tones.

Among social media sites, MySpace experienced a surge of popularity, followed by a sharp setback and then reinvention as an entertainment site at a lower level of usage. Many apps have been fads; e.g., the video game app Angry Birds was very popular for a while and then faded. Among the many other apps that have risen to great heights only to fall are Peach, Meerkat, Ello and Secret, names that most people have now forgotten but that had their moments.

Too often, when technologies fail, the companies that introduced them do not learn from the failure. There is a tendency to simply move on, throwing out research that indicated why it failed and laying off staff who had direct experience in developing and marketing the product or service.

1.5 Declines

If the pace of technology introductions has accelerated, it is reasonable to expect that the pace of technology declines (devices and services being replaced by newer technologies and services) would also accelerate. This appears to be the case. For example, landlines have been declining across Europe, North America and Asia and are being replaced by cell phones. This is particularly true among those 24-30. This also raises the issue of age versus generational differences. In the 20th century, young people did not have landlines. They used the landline in their parents' home, a public phone booth or perhaps a phone assigned to them in a dorm room. Later, as they moved out of the house, started a career and had their own home, they got a landline. As they aged, they changed their behaviour. Young people today do not have landlines, and they are not likely to get one as they age. It is a generational difference.

The list of both technologies and services that have declined in the past decade is quite long. Among the technologies that have declined are desktop PCs, DVD players, e-readers, digital cameras and non-smart cell phones. Among the services that have declined are snail mail, print telephone directories, DVD rentals, print newspapers and browsers on phones. These lists are not exhaustive. Depending on the time frame covered, PDAs, public phones, hotel phones and many early websites could be added to the list.

1.6 Advantages and disadvantages of being first

The question of whether it is better to be first to market or second was a conundrum in the 20th century, and it persists in this century. The potential advantage of being first is to attract general media attention, social media buzz and early adopters who can pave the way for a second wave of adopters. It may also help in striking deals with content providers who can learn first-hand what the technology or service can do. These advantages are mitigated by higher cost, the potential of software bugs, early negative publicity and a scarcity of content. With a second-in strategy, the product or service can probably come in at a lower cost, have more content and avoid some of the mistakes of the early entrant. As with many 20th-century technologies, it is hard to predict which is the better strategy. In general, a first-in strategy has the advantage if all the pieces necessary to succeed are in place: the technology works; it can be offered at a price early adopters will pay; there is sufficient content, if needed; and regulations do not impede it from entering the market. If one or more key components are missing or weak, it is often more advantageous to be enter the market second, when the missing piece(s) can be filled in.

The Internet of things provides a good example to illustrate the first-in or second-in advantages. An early entrant could pick off the low-hanging fruit, among thousands of possibilities, that seem to make a lot of sense (e.g., thermostats) and leave less likely successes (refrigerators that measure milk consumption?) to later entrants. However, a later entrant could assess the field and develop a more comprehensive strategy that is not apparent to early entrants.

There is another important factor in adoption that can play out in both first-in and second-in strategies: serendipity. Serendipity involves things that the company introducing the product or service did not plan. It happens sometimes because of luck—e.g., a major sports personality adopts the product and is seen widely using it, influencing others—and sometimes because a third party creates a service that piggybacks on the product or service and this leads to unplanned and unexpected new usage and users. A good example of this in the 20th century is mom-and-pop videocassette rental shops that emerged spontaneously in local neighbourhoods, allowing people to rent movies and providing a reason for many households to buy a VCR. A good example of serendipity in the 21st century is the hashtag used to group topics on social media. The first hashtag was proposed by Chris Messina in 2007 as a way to group topics on Twitter (Parker, 2011). Previously, it was very difficult to find related topics. Twitter did not adopt the hashtag at first, but the practice took off with users on Twitter and then other social media sites. It wasn't until 2009 that Twitter began to hyperlink all hashtags.

Ironically, some companies or industries have ignored or even tried to block serendipitous innovations. In the 20th century, the movie industry took mom-and-pop videocassette rental shops to court to try to stop them. The movie industry lost the case and as a result made tens of billions of dollars from videocassette rentals. In Twitter's case, the company simply failed to see the value of the innovation that was handed to it. This innovation, too, provided a significant revenue boost for Twitter, since it increased usage of and satisfaction with the service.

1.7 Media consumption, audience measurement and advertising revenue

The adoption of new technologies has affected how people consume media, especially video, which takes up most of the bandwidth on the Web. These changes, in turn, have affected audience measurement and the prices set for advertising on the Web and TV. We've come a long way since the 1950s when households had one device to consume video (a TV set), were given few content choices and often watched as a group. Audience measurement was much easier then because there were fewer variables.

Three characteristics of media consumption in the digital era complicate the process of audience measurement: device shifting, time shifting and place shifting. There are multiple devices for consuming media—e.g., TV sets, PCs, laptops, tablets and smartphones. Some of these were introduced only a few years ago; e.g., the iPad was introduced in 2010. The experience using these devices can be quite different. For example, the screen size could be 3 inches, 40 inches or even 70 inches. The device might stand alone, as in the case of a smartphone, or it could link Web video to a TV set; for example, hardware such as Apple TV, Chromecast and Roku, as well as videogame consoles and laptops, can stream video content such as Netflix or Hulu to the TV. This is sometimes called OTT or Over-The-Top. These devices may be used alone or with other devices in what is often called multitasking—e.g., using a smartphone while watching TV. Use of the two devices may be unrelated, as in sending emails while watching TV, or their use may be related, as in going on the website for a TV programme while watching that programme on TV. The latter is often called sync-to-broadcast.

Some device shifting provides a much better experience than a few years ago. For example, recent smartphones have much better resolution than the first generation of smartphones. Further, batteries are more powerful, consumers are used to recharging them constantly (many have chargers for the home, car and work) and they have much more access to Wi-Fi, which does not eat up a person's data plan. So the experience is qualitatively better, but this is not factored into audience measurement.

Time shifting has been around since the days of the VCR, but the VCR required physical hardware to record on (videocassettes), and it was difficult to program multiple recordings. The DVR made the process much easier, so much so that many people watch most of their TV programmes time shifted. Streaming services such as Netflix also provide content that is not locked into a rigid schedule. Nearly all video on websites is on demand, but it may not be available for an unlimited amount of time. Time shifting has led to video snacking (watching short clips on YouTube or Facebook) and marathon viewing (for example, when a person watches a few or several episodes of a series at one time, often on a weekend).

Place shifting involves the use of portable devices such as smartphones and tablets to access media just about anywhere. Place shifting also occurs in homes, as when a person watches a video on a tablet in any room. Often, there are patterns of movement—e.g., watching a video on a smartphone in the kitchen in the morning, in the living room in the evening and in bed at night. The potential opportunity for advertisers with place shifting is to provide location-based commercials—for example, a commercial for beer while the person is in a bar or a commercial for travel destinations at an airport. This is sometimes called dynamic ad insertion. It is more common on the Web, but it is under development by many TV service providers.

These patterns of media use have implications for audience measurement. They affect how people consume media (e.g., directly or indirectly, alone or with others), how immersive the experience is and what the degree of engagement with content is (Yang and Coffey, 2014, 55-60). Currently, advertisers know how many people watched a show or Web video, for how long and the demographics of viewers; advertisers know more about Web video viewers— for example, what else they do on the Web, what they have searched for and what they have purchased (Gunzerath, 2012). Many, especially television advertisers, want to know more about viewers and to be able to customise commercials based on this greater knowledge. They also want cross-media measurement—i.e., knowing what people do both in television watching and in Web use. Cross-media measurement is at an early stage of development. Some question the accuracy and reliability of audience measurements in these more complex environments (Napoli, 2014). At the same time, there is a reluctance to question too loudly these audience measurement numbers, since they are the coinage of the realm in determining the pricing of over $100 billion in advertising within the U.S. alone.

1.8 Social media and adoption

Research on the diffusion of innovations has found that early in the launch of new media, external influences such as advertising and marketing have more impact than internal influences such as word of mouth. Later, when many people have the technology or service, internal influences such as word of mouth have more impact (Rogers, 1995, 121-124). In the 20th century, this made complete sense. If there were 1,000 users of a new technology or service, how many people could they reach by word of mouth? So advertising was likely to be more influential. However, if there were 10 million users, word of mouth could be very influential, more so than advertising.

In a 21st-century context, social media has changed things. The scale and scope of social media are immense, and the chances of 1,000 active and vocal users of a new media technology or service influencing others are much greater. Whether through product reviews, Facebook posts, technology forums, group emails or other forms of social media, a relatively small number of people can reach a large audience. Further, many people who are considering the purchase of a new technology or service actively seek out reviews and comments by those who already have it. Traditional advertising still plays a role in early phases of adoption—but less so than in the past. Further, many advertisers understand the changes in the media environment and are themselves taking to social media to promote their products and services.

Social media has also become important in assessing how people react to video content on TV and the Web. These assessments are called social media analytics. As social media grew in volume and many companies began to pay attention to what people were saying, a number of groups emerged to quantify those comments and interpret their significance. They measure how many comments are posted on sites like Twitter and whether they are positive or negative. While useful, there are many limitations with the current state of social media analytics. For example, one programme may generate more comments than another, but this may not indicate that it is better liked. It could simply lend itself to gossip about the characters in the show.

Many of the specific methods social media analytics companies use are not published, since they are considered proprietary. This may be less of an issue in measuring volume, but the interpretation of comments as positive or negative is subject to the vagaries of linguistic patterns (a seemingly negative comment can be positive and vice versa, something readily distinguishable by humans but not necessarily by software code). It would be useful to know exactly how positive and negative comments are measured.

There is no dispute that some programmes generate a large volume of comments about them, but we know little about how representative of the general audience those commenters are. It is clear that some people post many comments, so the actual number of people commenting is less than the volume of comments. There are undoubtedly many lurkers who read but do not comment; their views are not known. There is also a dark side to social media comments—a tendency for those with negative views to air them vociferously (Streitfeld, 2013). This is a general pattern with social media comments, not just those made about videos and programmes. There is even a question as to whether all the commenters have watched a show or video. We know that in other areas of online comments, fake reviews are common (Streitfeld, 2012).

1.9 Privacy

We have come to accept that websites record what news stories we read, what videos we watch, what we buy online and what terms we use in search engines like Google. So if we search for dining room chairs on a search engine, the next day we will see ads for dining room chairs on multiple sites. Similarly, if we watch a video of Adele on YouTube, the next time we visit YouTube, the opening screen will be filled with videos of Adele and singers with styles similar to hers. In some ways, this can be seen as positive: isn't it better to see ads for products we want and be shown content that interests us (Lohr, 2015)? However, this comes at the cost of losing some of our privacy.

As technologies like voice recognition and miniature video cameras become more advanced and cheaper, new privacy concerns are emerging. One is that voice recognition systems can record our queries or even what is said in the room that is unrelated to media use; another is that miniature cameras built into devices might observe what we are doing in a room. Verizon has filed for a patent on a technology that can be embedded in a

DVR and watch as well as listen to what is happening in a room (Huffington Post, 2012). Its intended use is to deliver targeted ads, but its wider applications are clear and foreboding.

1.10 Conclusion

We are in a period of rapid technological change and the proliferation of many new media technologies and services. At the same time, there has been a remarkable consolidation of content and applications in one device—the smartphone. There is really no historical precedent for this. The personal computer may be the closest example from the 20th century, but very few PC owners in the 20th century had anywhere near the number of applications that are in smartphones today.

Social media has thrived on user-generated content. However, user-generated content will not be sufficient for many high-resolution technologies such as VR and 4K TVs. Some existing content (e.g., movies) can be adapted, but much will have to be produced. The business model for this—cost to produce and potential revenue—is a work in progress.

This chapter has also shown that a historical perspective on adoption of new media is valuable to the study of recent media adoption. It can help us measure the pace of change today in comparison to earlier periods, avoid pitfalls of earlier new media, learn positive lessons about marketing successful technologies and check in the dustbin of failed technologies for what might be a false start. There are no perfect analogues in the past for a new technology today, but there are lessons. A quote attributed to Mark Twain (it's not clear if he actually said this) sums it up: 'History doesn't repeat itself but it does rhyme.'

References

Brody, Y. (2013, May 7). Losing Sleep in the 21st Century. *Psychology Today.* https://www.psychologytoday.com/blog/limitless/201305/losing-sleep-in-the-21st-century. Accessed 2 January 2017.

Carey, J., & Elton, M. (2010). *When Media Are New: Understanding the Dynamics of New Media Adoption and Use.* Ann Arbor: University of Michigan Press.

Dutton, W. (Ed.). (2013). *The Oxford Handbook of Internet Studies.* Oxford: Oxford University Press.

eMarketer. (2016, June 6). Growth in Time Spent with Media Is Slowing. https://www.emarketer.com/Article/Growth-Time-Spent-with-Media-Slowing/1014042. Accessed 4 January 2017.

Gunzerath, D. (2012). Current Trends in U.S. Media Measurement Methods. *International Journal on Media Management, 14*(2), 99-106.

Herrman, J. (2016, April 17). Media Websites Battle Faltering Ad Revenue. *New York Times,* p. C-1.

Huffington Post. (2012, December 5). Verizon Files Patent for DVR That Watches Viewers. http://www.huffingtonpost.com/2012/12/05/verizon-dvr-patent-spies-tv-advertising_n_2246973.html. Accessed 20 November 2016.

Katz, J. (2006). *Magic in the Air: Mobile Communication and the Transformation of Social Life*. New Brunswick, NJ: Transaction.

Lohr, S. (2015, October 5). With the TV Business in Upheaval, Targeted Ads Offer Hope. *New York Times*, p. F-6.

Napoli, P. (2014). The Local People Meter, the Portable People Meter, and the Unsettled Law and Policy of Audience Measurement in the U.S. In J. Bourdon & C. Meadel (Eds.), *Television Audiences Across the World: Deconstructing the Ratings Machine* (pp. 216-237). New York: Palgrave Macmillan.

Oxford Internet Institute. (n.d.). www.oii.ox.ac.uk. Accessed 20 April 2018.

Parker, A. (2011, June 10). Twitter's Secret Handshake. *New York Times*, p. ST-1.

Pew Research Center. (n.d.). Internet and Technology. http://www.pewinternet.org/2018. Accessed 15 April 2018.

Pew Research Center. (2015, October 29). The Demographics of Device Ownership. www.pewinternet.org/2015/10/29/the-demographics-of-device-ownership/. Accessed 1 April 2016.

Rogers, E. (1995). *Diffusion of Innovations* (4th ed.). New York: Free Press.

Sterling, C., & Haight, T. (1978). *The Mass Media: Aspen Institute Guide to Communication Industry Trends*. New York: Praeger.

Streitfeld, D. (2012, October 18). Buy Reviews on Yelp, Get Black Marks. *New York Times*, p. B-1.

Streitfeld, D. (2013, July 15). Why Reviews on the Web Say Bad Things. *New York Times*, p. B-1.

Williams, E. (1978). *Research at the Communication Studies Group*. London: Post Office Long Range Studies Divisions.

Yang, Y., & Coffey, A. (2014). Audience Valuation in the New Media Era: Interactivity, Online Engagement and Electronic Word-of-Mouth. *International Journal of Media Management, 16*(2), 55-75.

2 Change of society by globalization: The intercultural and sociocultural impact of globalization on individual sectors

Mike Friedrichsen

2.1 Introduction

New technologies and global media are affecting societies and political structures.

This hypothesis is guiding this chapter. The main focus lies on the impact of the global media industry, new technologies and social networks on individual sectors in terms of intercultural and sociocultural tensions. Technology companies enter previously existing sectors due to new technologies such as the Connected Car. With the current development of the Google Car, the technology giant Google enters a new market. By developing self-driving cars, Google will compete in the future with German automobile manufacturers like Mercedes and or BMW. It is not just Google entering the future automotive market as new competitor but also Apple and the start-up company Tesla Motors. These innovative and financially strong competitors will have a direct impact on German society, provided that changes are possible in the labour market or in German laws (Abdi, 2014).

Social media and international media groups have a global effect on political structures in different countries. In countries such as China, Tunisia and Russia, freedom of speech and freedom of the press are limited. Due to the social platforms and the global reporting, the pressure will increase on the existing governments. The west-mediated fundamental values are in contrast to the predominant values of different cultures (cf. GF, 2015).

In the face of the hypothesis 'New technologies and global media are affecting societies and political structures', the following research question appears: How far are new technologies and media globalisation influencing the sociocultural structure, particularly through social media, in selected countries like Germany and Russia at present and in the future?

In order to explore these effects on the sociocultural sector, this chapter examines individual cases in the respective countries on the basis of statistics and reports. For this purpose, this chapter examines the technological impact of globalisation due to the new market entry of the technology companies.

In a further step, reference is made to the influence of social media on political and social movements in the tense political situation in the countries.

Furthermore, this chapter will study the impact of global media groups on culture groups. 'With [the] help of products of the modern entertainment industry, movies and television are publicizing the same ideas of an equal life all over the world, western values are increasing in different culture groups. A legitimate question is, if it's generally legitimate to convert other countries and cultures to value and ethics of western civilization' (University of Media Stuttgart, 2015). Afterward, the results will be gathered in one collective context to finally show a preview of intercultural and sociocultural changes in the future by technology industries and media groups. The aim of this chapter is a neutral elaboration on the named main focuses. A neutral evaluation is important for our result, which is an objective look at the impact of the western-shaped, global influence of technology industries and media firms on other cultures and ventures.

2.2 Definitions

Globalisation: Globalisation is the increasing worldwide network of nations in all areas such as politics, economics, communication and culture. This happens both between individuals and between organizations, enterprises, societies and states. Globalisation has been driven mainly by the advances in communication and transportation technologies and affects the contact intensity between members of different cultures (cf. IKUD, 2009). Other designations are *mondialisation* (from French) and *denationalisation*. In the latter term, the loss of power and the meaning of the individual nation state are emphasised (cf. GF, 2016b). Cultural globalisation states that it also reaches to the level of cultural ideas, designs and identities in increasing mutual interconnections and influences (cf. IKUD, 2009). A nation is characterised by a common culture and language and also often by religion. In times of global exchanges, the boundaries are becoming increasingly blurred by people's massive movement and unrestricted freedom to travel. Most of the effects are perceived as positive enrichment. Thus, people can find restaurants with exotic foods all over the world, and numerous spectators can find on stages all music and art from around the world. Often, the influences of the leading, most western, industrialised nations are also critically evaluated. Smaller cultures are often overwhelmed by western flair, their own language and centuries-old traditions are losing importance and a cultural mainstream expands. In this respect, many scientists find that the driving aim is not the unification but the co-existence of different cultural influences. Cultural diversity is an expression of the globalised world, as well as the increase of international decisions in politics. The globalised policy is necessary because particular problems, such as environmental protection or terrorism, cannot be fought at the national level. Only through international cooperation and multilateral principles can these problems be addressed (cf. GF, 2016b).

Digitalisation: There are two interpretations for digitalisation, digitalisation as transfer and digitalisation as process (cf. Hess, 2013). Digitalisation as transfer describes the digitising of analogue media to binary formats. The transfer of analogue media to digital media has enabled new media forms like social media and convergences between traditional media forms. A new interpretation of digitalisation defines the relationship between digitalisation and the individual, organisation or society (ibid.). Digitalisation as process focuses on changes in behaviour and environment caused by digital media. As an example of the effect of digitalisation at the individual and private level, social media changed human relationships between contacts. For organisations, CRM and ERP systems are changing the whole environment and adding opportunities like networks—internal and external. Internal networks allow fast communication between all departments, and cloud systems enable collective working on digital documents or simple EDI. External networks support communication between different organisations. Companies easily order products and resources from suppliers by SCM without delays in the value chain. Societies benefit from new infrastructures and more participation in political, commercial and social problems. For example, educational systems are changing because e-learning and new technologies support education.

Autonomous connected car: A connected car is a vehicle 'that provides Internet access to all the mobile devices used by the driver and passengers. It accesses the Internet via cellular or satellite communications and provides tablet-sized screens for passengers or a Wi-Fi hotspot for passengers' own devices' (pcmag, 2016). Connected cars also allow the use of apps like parking apps via smartphones (cf. acc, 2016). The usage of apps and new sensors simplifies navigation and reduces accidents. Because of this networked car, automobile diagnostics are enabled anywhere and everywhere as long as Internet connectivity exists. Drivers can react in time if serious problems appear. There are two different communication systems in connected cars (ibid.). First is the V2V system, which enables communication between vehicles. With the V2V system, drivers can communicate with each other, and each car can detect other cars via sensor. Second is the V2I system, which allows interactions with the car's environment. V2I is capable of 'stopping a car from entering a dangerous intersection' (acc, 2016). Both are created for the safety of road users. In contrast to connected cars, autonomous connected cars are autonomous driving cars. These cars are not driven by humans; systems will lead autonomous connected cars to selected targets.

Social media: Social media allow the networking of users and their communication and collaboration via the Internet (cf. Bendel, 2016). The user can interact communicatively via both messages and comments, as well as actively participate in the design of media content on a specific platform on the Internet. This user-generated content clearly differentiates social media from traditional mass media (print, film, radio and television) because social media dialogues arise from media monologues. The user—who so far has been only a consumer—becomes himself a producer because of the broad opportunities and low entry barriers of social media. Typical social media applications are weblogs such as Spreeblick, micro-blogging services such as Twitter, social networks like Facebook and StudiVZ, recommendation sites like Qype, photo-sharing sites like Flickr and video-sharing platforms like YouTube (cf. Gründerszene, 2016).

2.3 Intercultural effects of globalisation

Globalisation and new media are changing society (cf. Stumberger, 2012). Nowadays, new technologies and social media arise quickly and control our everyday life.

2.3.1 *Effects of new technology on the German car industry and sociocultural structures*

Germany as an industrial country is an example of technological influences on existing markets. Within globalisation, Germany is one of the biggest export nations. The next section will show why Germany is so successful in exporting goods, which sectors will be in danger from present technological developments and what effects this could have on Germany's society.

2.3.1.1 *'Made in Germany' as a standard for high quality and values*

'Made in Germany' was established as a warning label in 1887 because of the bad quality of German products (cf. Wied, 2016). After some years, German products were being improved, and 'Made in Germany' became a label for high quality. 'Made in Germany' is associated with well-paid employees, high quality and high reliability and innovations (ibid.). Nowadays, 'Made in Germany' is a seal of quality for German products and one reason for Germany's wealth. Germany's companies earned €1.133,6 billion by exporting German products (cf. Spiegel, 2015). This profit out of exported products was a new record in Germany, and after China and the U.S., Germany is one of the biggest export nations worldwide (cf. Statista, 2015a). One reason for this high revenue is German values and high quality associated with 'Made in Germany' (cf. Stocker, 2014).

But 'Made in Germany' is in danger (cf. Slavik, 2013). Customers are paying attention to labels from organisations to make sure that products are produced in human- and eco-friendly environments (ibid.). Scandals like the VW diesel scandal are influencing 'Made in Germany', and many car consumers have lost trust in German values like honesty (cf. Handelsblatt, 2015). Because of new technology, the car manufacturer VW was able to manipulate its cars' emissions. So VW advertised low car emissions and cheated its customers concerning an emissions test in the U.S. (cf. Hotten, 2015). With more than 600,000 jobs, VW is one of the most important employers in Germany (cf. Welt, 2015a). Thus, as a result of this scandal, VW will get a high penalty, and this will also have effects on VW's employees and the label 'Made in Germany'.

2.3.1.2   *New Competitors—Google, Apple and Tesla Motors*

Related to VW, German car manufacturers like Audi, BMW and Mercedes are important for Germany's wealth and growth (cf. Bundesregierung, 2016). With about €250 billion in revenue and one-fifth of the total revenue from German industry, the car industry is one of the important industries in Germany (ibid.). The revenue of the German car industry was €368 billion, with 5.91 million cars produced in 2016 (Statista, 2016b); Germany is behind only China (23.7 million cars), U.S. (11.66 million cars) and Japan (9.77 million cars) as one of the biggest car producers in 2016.

Technology companies like Google and Apple are now entering the U.S. car industry with innovations like connected cars and autonomous driving cars. Furthermore, Tesla Motors, as a new electric car manufacturer, has also enriched the U.S. car industry. In Germany, 32% of car drivers could imagine buying a car from Google, Tesla Motors or Apple (cf. Statista, 2015c). These technology companies are more favoured in expanding markets like India and Brazil. In Brazil, 63% of car drivers could imagine buying cars from technology companies (ibid.). A great many potential car buyers in China (74%) and India (81%) would prefer Google, Apple or Tesla Motors cars instead of Toyota, VW or GM cars.

Google, as an example, is developing Google car to enter the car industry. Google car, as an autonomous connected car, will be navigated via Google Maps (cf. BDI, 2015, p. 21). Development and testing of Google Car will prepare this new product for the mass market (ibid.). Google is also involved in medical, robotic and drone projects. With new drones, Google is trying to transport radio signals and develop new supply possibilities. In medicine, Google is investing in the development of lenses that allow automatic measuring of blood sugar levels.

Apple as a tech company wants to equip each car with a new Apple iPad (Pallenberg, 2015). Within rumours of an iCar, Apple wants to be included in connected-car technologies to reach 1.2 billion potential new customers (ibid.). Apple is about to produce iPads that can be used in every vehicle. Apple has already invented Apple CarPlay, which is a feature that allows the use of Apple apps in vehicles by connecting iPhones with onboard radio systems or similar technology (Wiesmüller, 2014). With CarPlay, drivers are able to use a navigation system app, listen to music from of iTunes or make phone calls while driving. Besides rumours about Apple's iCar, Apple is headhunting professionals of Tesla Motors and Mercedes and has new electric motors in development (cf. Schmidt, 2015). Apple CEO Tim Cook mentioned that software will be more important for vehicles and that autonomous connected cars are an important aspect of future mobility (ibid.).

Tesla Motors, as a manufacturer of electric cars, started to sell the Tesla Roadster as its first product in 2008 (cf. Statista, 2016a). Revenue of Tesla Motors were US$3 billion in 2014. The U.S., Norway and China are the biggest markets for Tesla Motors so far (ibid.). In Germany, there were 800 Tesla cars licensed in 2014. Worldwide Tesla sold 35,000 cars in 2014. In 2015, Tesla revenue was US$1.2 billion, and the company loss was US$320 million (cf. Eckl-Dorna, 2016). Stock in Tesla rose to $US260 a share in July 2015 (cf. Bourse, 2016). In February 2016, the stock dropped to US$140 (ibid.). Within fluctuating income, Tesla wants to sell up to 80% more cars in 2016 (cf. Eckl-Dorna, 2016).

2.3.1.3 *Resulting consequences for the German car industry and sociocultural structures*

There is enormous potential for Germany and Europe in the existing chain of values (cf. BDI, 2015, p. 3). In Europe, growth of gross value could be €1.25 billion through 2015 (ibid.). To reach this growth, companies have to support digitalisation on an organisational level and in society and to invest in new products and technologies (ibid.). Furthermore, European standard-isation of industry is important for the next years (cf. BDI, 2015, p. 14). On the other hand, a loss of €605 billion is possible. For Germany, there is a chance of a €220 billion loss (cf. BDI, 2015, p. 10).

Germany has been one of the top industrial nations for decades now (cf. BDI, 2015, p. 9). Industry made up 22% of Germany's economic performance in 2013 (ibid.). In contrast, industry make up 12% of U.S. economic performance. The important sector in Germany's industry consists of global car manufacturers like BMW, Mercedes and VW. Germany is addicted to industry and especially the car industry.

The next sections will form an estimate of the effect of negative and positive scenarios on Germany's industry and sociocultural structures.

2.3.1.3.1 NEGATIVE SCENARIO

In case of negative growth, because of the damaged quality label 'Made in Germany' and new strong competitors in car industry such as by Google, Apple and Tesla Motors, effects on Germany's society could be enormous.

In January 2016, the unemployment rate was 6.7% in Germany (cf. destatis, 2016). Because of the flow of refugees, Germany's Secretary of State of Employment Nahles declared rising unemployment rates (cf. Welt, 2015b). If Germany's car industry can't keep up with its new competitors in the next few years, more jobs are in danger. In the worst case, Germany's whole economy will face recession, and the unemployment rate will exceed 10% percent. Because of missing revenue from the car industry, wealth in Germany will sink. Changes in society and politics will be the consequence of rising dissatisfaction. As a result of the refugee crisis, part of German society will shift to the right because of dissatisfaction with German politics and fear of refugees (cf. Lühmann, 2016). Rising dissatisfaction and economic crisis can cause formation of radical groups, and disturbances can arise. Germany's role as export leader and affluent country will disappear, and the European force driving the economy will stop.

2.3.1.3.2 POSITIVE SCENARIO

German car manufacturer VW and European manufacturers like Fiat and Renault are members in open automotive alliance led by Google (cf. BDI, 2015, p. 13). Open Automotive Alliance is a platform for collective development of connected car systems (ibid.). With alliances between manufacturers and standards in the European Union, Germany's industry will profit from new technologies. New competitors like Google are entering the market, but through OAA, VW will also profit from connected cars. With investment in new technologies, VW, Mercedes and BMW will participate in future markets. New jobs will arise, and the economies in Germany and Europe will become stable or rise. The unemployment rate will sink, and citizens in Germany and/or Europe will be more satisfied with liberal and social politics. With investment in new technologies like electric motors, green energy will replace fossil fuel. Eco-friendly driving will help in fighting the problem of particulate matter in cities like Stuttgart. Driving will be a new experience because of autonomous connected cars. The rate of accidents will sink in Germany, and traffic jams will be prevented. With alliances and investments in new technologies that lead to a more affluent Germany, future problems in society and the integration of refugees could be solved more easily.

2.3.2   *Cultural globalisation—Adjustment by social networks*

With social media, we can exchange experiences among individuals or employees. You communicate and you work together by using text, image and sound (cf. Bendel, 2016).

Social media serve for both direct personal and professional communication, as well as the dissemination and discussion of informational and entertaining content. The different intensities of communication reach from private, one-to-one contact on the exchange different groups up to communication with all users of a network (cf. bpb, 2010a).

The use of social media is part of everyday life for many people. The following section deals with the question of what impact social media have on society, culture and politics.

2.3.2.1   *Influence of social networks on society and culture*

If one thinks today of social networks, Facebook, Twitter, Instagram and further widespread communication possibilities come up to mind. For many people, the communication via these platforms is almost indispensable and has become a big part of everyday life. Chatting, playing and the exchange of data are characteristic for social media. In particular, they provide the opportunity for users to present themselves using a personal profile. So we often divulge much of our personal data, which other users can view on the Internet (cf. IuG, 2010).

In the ranking of the largest social networks and messengers in 2015, listed by the number of active monthly users, Facebook is in first place. The micro-blogging platforms Twitter, Google+ and Instagram follow Facebook at some distance. In the second quarter of 2015, Facebook had nearly 1.5 billion monthly active users worldwide. In Germany, with more than 600 million visits in December 2014, Facebook is the most popular of the social media (cf. Statista, 2016b). In the fourth quarter of 2015, Internet users from all over the world spent an average of 1.1 hours a day on social media (cf. Statista, 2016c). These are huge numbers, which might explain part of the real impact of social media on people's everyday lives today.

Thus, the rapid development of information technology and the associated new digital communities and networks has had an immense impact on the way we communicate today. With the introduction of the Web 2.0, our possibilities for online-based identity acquisition, relationship acquisition and information acquisition have increased. This development can not only lead to a negative influence on today's society and culture but also have a positive influence.

Communications expert Robert Spengler believes that social media can have a negative impact on society because direct communication becomes stunted by them more and more. Artificial communication facilitates our everyday lives. But important skills can become stunted because many people then forget how human communication actually works. People replace encounters in the real world with virtual contacts. They tap more and faster than ever—but also incorrectly, superficially and impersonally. Consequently, they talk less and less. Above all, digital stress can often miss the real thing in life: real human encounters (cf. Spengler, 2012).

This development is also shown among the younger generation, which does not know life without the Internet and its associated constant accessibility. The impact of the Internet and social media has been most significant in the way it has changed the lives of young people. Their digital children's room shows a childhood that is significantly affected by social media, mobile phones and the Internet. Friendship interaction and face-to-face communication are increasingly conducted online or through text messaging. These interactions have had major cultural consequences. Online communications have influenced the evolution of language. They have thrown up new rituals and symbols and have had an important influence on people's identity. Consequently, young people do not know how to hold a conversation offline, how to behave in such face-to-face situations and how to present themselves in real life. Texting and further online communication forms, such as symbols and pictures, constitute the immediate impact of social media on the everyday culture (cf. Ahmad, 2015). All in all, social media can lead to less face-to-face communication and to less personal encounters, which also leads to a loss of the ability to approach people, hold a conversation or win sympathies.

But, on the other hand, social media can also have a positive influence on society and culture, as social media enable people to feel as part of a community that shares the same fundamental interests and values and gives them the opportunity to exchange information and to participate in conversations and in decisions. Basically, social media are primarily a platform for the emancipation of the user. This emancipation has a direct, definite consequence for politics, society, economics and marketing. The interesting thing is that people are becoming more familiar with the world of social networks across the world. If these people then have to report something important, they will also increasingly use these networks. They will retain less information for themselves and participate more in discussions. One country after the other will see how the exchange of information via social media has changed radically. Each message that generates enthusiasm has also the potential for viral distribution. Today, they are primarily messages about parties or gossip. Now, the customers themselves act as multipliers. If they want to assess or comment on a product or service, they can spread this information quickly and easily via social platforms. If they are dissatisfied, they can manifest this displeasure virally. The more people have access to smartphones, the more quickly this development proceeds. People are no longer bound to a place and are everywhere and always connected and reachable (cf. Kirkpatrick, 2011).

Someone who is logged on to a profile is also at the same time part of a large online community and thus gets a certain feeling of togetherness. The *Piratenpartei Deutschland* describes the digitised society as supportive of progress because personal opinions that are represented on platforms can sometimes have an impact on decisions. These opinions are also reflected outside the media world. This is, for example, demonstrated by the protests in relation to the Arab Spring in 2011 (cf. Damrow, 2012). Consequently, social media can influence political structures by supporting freedom of opinion, in which the users can exchange their personal opinions and organise social movements to fight against prevailing political structures.

2.3.2.2 *Effect of social networks on people's participation in political subjects*

Social media have a huge impact on public opinion and people's actions in relation to politics. The Arab Spring in 2011 is an example of the enormous influence that social media have on politics. In the spring of 2011, hundreds of thousands of people in the Arab world took to the streets and protested against autocratic regimes, which was called a quick Facebook revolution.

Despite censorship and repression, social media helped the dissatisfied people to organise demonstrations and to defy the state power. Thanks to Facebook, Twitter und YouTube, the protest made it from the living room to the street. The citizens learnt via social media that they were not alone with their anger and that other people also dared to raise their voices. Arabic countries affected were Egypt, Tunisia and Syria. Thanks to this 'technological revolution', critical mass has come together, said researcher Fadi Salem of the Dubai School of Government. Research results show the importance of Facebook, Twitter and YouTube. During the riots, a team led by Tufekci interviewed 1,000 users in Egypt. Although the survey is not representative, the results provide a good insight: almost half of the surveyed Egyptians (48%) learnt about the protests in direct talks, in addition at least 28% on Facebook. And a quarter of the Egyptians disseminated on their own several pictures and videos of the demonstrations in Tahrir Square and elsewhere. In a large-scale survey of Arab Internet users, Fadi Salem found out that social platforms were used less for private chats and more for political purposes during the protests. Up to 90% of Tunisian and Egyptian Facebook users had used Facebook to disseminate information. The social networks had a massive influence on the Arab Spring—not only among the educated but also among those in all milieus. This shows that social media are especially used for exchanging views on political subjects in countries where dissatisfaction and repression of freedom of opinion dominate (cf. Focus, 2012).

With Internet censorship in their countries, the governments of China and Russia are trying to prevent protests via social media, like the protests of the Arab Spring. Any comment or any platform that represents a threat to the government is blocked. These primarily include social media, like Facebook, Twitter and YouTube. In China, there live 1.3 billion people, of whom 600 million have Internet access, according to the CNNIC. While the Chinese firewall blocks foreign sites like Facebook, domestic Chinese social media like Weibo are offered. However, these social media are monitored by the government. The Communist Party of China wants to control every public expression exactly. Critical content will be deleted in order to prevent potential unrest in Chinese society (cf. Scheuer, 2013). Also in Russia, conservative Russian politicians want stricter control over the Internet. A big demonstration by a Facebook group is not possible in Russia because Internet sites that promote calls to unauthorized gatherings can be blocked without a court order. In addition, bloggers are more strictly controlled. Each user of a blog or an account on social networks, like Twitter, Facebook or Vkontakte, is entered in a register when its page is visited by more than 3,000 readers per day. It can be assumed that this law is an opportunity for the government to put bloggers under pressure and finally to close their sites (cf. Smirnova, 2014). This suggests that free opinion formation via social media is not possible everywhere and can be strictly controlled by governments in order to prevent unrest and possible protests and to impede democratisation. However, people are familiar with methods that enable them to get access to blocked websites. For example, it is assumed that most Chinese know how to bypass the Internet censorship in order to get access to social networks such as Facebook (cf. Welt, 2010).

Social media also pose dangers. Extremists can use them to spread their messages and to serve as propaganda platforms. One extremist group is the ISIS militia, which uses social media like Twitter and Facebook as a propaganda machine to spread its ideology and hate messages. At the end of 2014, a study by the Brookings Institution revealed that there were already tens of thousands of Twitter accounts that were attributable to the ISIS and its supporters. This service is a key propaganda platform for the terrorist group, and Twitter wants to take action against it. The short message service has already adapted its guidelines, and now threats or advertisements for terrorism are classified as abuse and consequently violate the terms of use (cf. Spiegel, 2016). Obvious propaganda of extremists can be prevented with this measure. However, this raises again the question of whether Twitter and other social media that act in such a way do not allow freedom of expression. To protect the security of society and to prevent further terrorist attacks, this limitation on the freedom of expression can be viewed as necessary and is accepted by the majority of society.

Another point is that there is also strong evidence that the sorting algorithms of social media contribute to opinion formation. One who gets to see only radical views and assertions on Facebook and elsewhere perhaps is more likely to become radical. Consequently, social media are presumably changing society. The current refugee crisis in Europe is a topic of conversation on social media for many people—including those in Germany. Unclear and unconfirmed allegations increasingly lead to distrust of refugees and therefore to a distorted picture of the reality that people might believe. In networks, rightist propagandists can use these sorting algorithms very effectively to target and spread fictitious rumours about refugees. Users in networks have a tendency to gather in communities of interest, so that they get to see especially appropriate content. This leads to 'confirmation bias, division and polarisation'. Confirmation bias is a psychological phenomenon, in which humans prefer to close knowledge gaps with allegations they already believe. This leads to the distribution of distorted narratives, fueled by unconfirmed rumours, distrust and paranoia (cf. Stöcker, 2016). Thus, social media can also lead to one-sided opinions, since users cannot see all the information and facts needed to form informed opinions. The one-sided flow of information may eventually lead to a reorientation of political views.

2.3.3 *Global media as motor for the cultural change process*

The development of the so-called information society via networked devices is the foundation of globalisation in the cultural and social field. The Internet, as well as all other globally provided media of modern mass communication such as satellite TV, combines the cultures and societies of the countries in the world across all political and geographical boundaries. Apart from a global understanding and a desirable exchange of world cultures, critics see also dangers in the new global availability and distribution of media content and information. It is feared that previously independent cultural traditions and social values will be aligned, especially due to the powerful influence of exported U.S. culture and therefore also U.S. ideology (cf. University of Media Stuttgart, 2015).

The products of the modern entertainment industry, such as music, film and television, propagate the same ideas and the same lifestyle worldwide. Western values reach foreign cultures increasingly. The question arises whether global media actually propagate western values and what influence global media generally have on society.

2.3.3.1 *Music, movies and television series as intermediaries of western cultural values*

The entertainment industry is considered an intermediary of lifestyles and living and plays therefore particularly a role for young people as an orientation for the formation and expression of their personal style.

Currently, there are four so-called major entertainment labels. Their share of the worldwide revenues is about 80%. The 'Big Four' are the Universal Music Group (U.S.), Sony Music Entertainment (Japan), EMI Music (UK) and the Warner Music Group (U.S.). These multinational corporations dominate the global music market and decide which songs will be played. The entrepreneurial oligopoly in the music market is in particular therefore criticized because the globally offered content mainly transports western values and ideologies. Among them is the English language, with specific and new terms. This leads to the fact that central areas of cultural life are influenced by universal imagery, mode shapes and consumer goods. Criticism of the monopolisation of the music market has led in Germany to the requirement of a government-regulated quota for the entertainment sector, such as in France, Switzerland, Canada or Poland. In France, for example, broadcasters have been obliged since 1996 to have French-language titles make up 40% of their content, and at least half of them should not be older than six months. The German Bundestag enacted a German music quota on 17 December 2004. With a commitment of German radio stations, a 35% share of music by German artists or produced in Germany is to be achieved. A final decision on a music quota can be made only by the respective provinces, since they are responsible for broadcasting policy. The aims of this controversial requirement are the prevention of a monoculture created by globalisation of the music market and the support of national and regional diversity (cf. bpb, 2010b). It becomes apparent that the government itself perceives an increasing influence of western values on songs that are disseminated through global media such as radio, television and the Internet. Therefore, the government tries to prevent, by the introduction of state regulations, the playing only of western music of the so-called Big Four. By these regulations, local-language titles, which reflect the culture of the country, are to be played and thus bring the local culture closer again.

As a mass medium, television has a far-reaching effect on its consumers. As television and films are easy to consume and are very popular in all populations, they play a central role in expression and style formation. For some time, entertainment programmes such as the talent show *Pop Idol* have achieved the highest audience ratings. Within the popular entertainment segment, international trade in specific programme formats has increased significantly in recent years. The integration of broadcasting formats leads to the increasing alignment of television programmes worldwide.

Due to the fact that communication technology and television have been developed and spread around the world, the idea of man and the conception of the world will be changed for many people by the content of television. Communist or authoritarian countries, such as North Korea, use the television as a way to disseminate propaganda. However, globalisation critics decry the dominance of U.S. films in the telecommunications channels. Critics, especially those from an Islamist background, see this as an insidious infiltration of western values, which they see as a threat to their culture and lifestyle.

In contrast, many proponents of globalisation welcome the influence of western values such as the equality of women, the legal system and freedom in religious matters. Especially in African or Muslim societies, where patriarchy still exists, these values could lead to more freedom and therefore to a higher standard of living (cf. GF, 2016a).

2.3.3.2 *Influence of global mass media on personal self-improvement*

After schools, media are the main source of lifelong learning and political education. This includes not only textbook knowledge of the structure of state institutions but also personal experiences with the actual processes of political decision-making. What powers the president has, according to the basic law, people learn in school. What role a politician actually plays can be understood only by observing daily events—consequently by using media.

Mass media influence people's opinions about politicians, political parties and social controversies. In the short term, the media coverage of parties or individuals often changes very quickly. Politicians who are criticized by the media can lose support among the population a short time later. Even elections can be decided in this way because shortly before the election more and more people are uncertain which party they should give their votes to. In the long term, the media coverage also influences the fundamental opinions of citizens about politics. Thus, since the early 1990s, the increasingly observed disenchantment with politics is attributed to the fact that the media criticize predominantly all parties in the long term and also describe politicians as causers rather than solvers of social problems (referred to as the media malaise hypothesis).

Mass media influence awareness. People can give their attention to a few social problems. But they mainly concentrate on those problems that are most frequently reported by the media (referred to as the agenda-setting effect). Therefore, they consider those problems as important. This effect can be positive, on the one hand, because it contributes to the fact that a society agrees on certain issues that need to be solved. However, on the other hand, this effect can have negative consequences if the media deal with issues that are actually secondary and distract from the real problems. Examples include the extensive media coverage of alleged risks like BSE or swine flu, which occupied Germans for months, although the actual threat was rather low. Mass media can therefore have many different effects. From a social perspective, some of them can be seen as positive, but others have to be viewed as negative. That they occur has nothing to do with the fact that people are gullible or manipulable. People often have to rely on the media reports if they want to make a judgement because they have no other source of information (cf. Bpb, 2011).

2.4 Result

New technologies and global media are affecting societies and political structures in the following ways. Advantages of new technologies like autonomous connected cars are a more comfortable lifestyle, more security and possibilities for eco-friendly driving. New technologies can create and also can delete jobs. Humans are connected with digitalisation at private, organisational and societal levels. Social networks enable human relations via the Internet and new possibilities to participate in social and political issues. Misbehaviour in politics and industry is now 'public' for everyone to see anywhere and anytime. If misbehaviour happens in politics, work or society, people will react with public criticism in social networks. And all over the world, people can hear and see what happens in other countries because you can consume your favorite media all around the world. Societies are no longer addicted to the content of media that government supports. Access to global media and new technology is all that is important, and people will generally use them. Consequently, social media assume an increasingly important role in people's everyday lives today.

Social media also affect society and culture both positively and negatively. On the one hand, social media strengthen the freedom of expression for people of different cultures and nations and promote the feeling of togetherness, as they can represent and share common interests on social media platforms. On the other hand, social media and the constant networking lead to a pressure to be always reachable. Consequently, real social communication skills, such as holding a conversation face-to-face, decrease due to the increasing communication via Internet. Especially those in the younger generation are affected by this development because they grow up using social media.

Global media, such as music, films and television, convey especially American/western values. In the music market, for example, they are conveyed through so-called major labels that publish mostly U.S. music titles. Therefore, European countries such as Germany use a state quota to prevent the playing of only American titles and to ensure titles produced in Germany are played on the radio. This should bring the local culture closer and prevent the loss of this culture. But the distribution of western values such as equality of women, the legal system and freedom in religious matters can also have a positive impact, particularly in African or Muslim societies. These values could lead to more freedom and therefore to a higher standard of living in these societies.

Within new technologies and global media, if companies or countries don't invest in new products and infrastructure, they can't keep up with innovative competitors all around the world. Structures in nations, societies and companies have to change to adapt to new conditions and environments. Politics and management have to react, or people will become dissatisfied and/or jobs and wealth will be lost. Demonstrations of freedom of expression have reached a new level through social media. If people don't want to be kept down by politicians or ripped off by big companies, they can use social media to form communities that take a stand against racism, suppression and exploitation. But, on other hand, extremist groups are also using social platforms to communicate and plan demonstrations.

This is the reason why it's important to learn responsible handling with each new gadget and opportunity via the Internet. Responsibility with social media and new technologies like smartphones should be taught as a new school subject to prepare young people for dangers that arise through these gadgets. Irresponsible handling of new media and social media can escalate quickly. Tirades with racist and defamatory comments can't be tolerated. Laws and security have to be prepared for contraventions to prevent irresponsible handling of new media. Each new medium or technology is not evil, but humans can do evil things with it if they don't know how to handle it in a responsible way.

At the least, new technologies and social media enable new ways of collective work and life. Beside new complex structures, dreams like a united European Union are more realistic than ever before. Through social networks and global media, people all over the world and especially in Europe are in contact with each other. The dream of liberty and democracy is expanding, and structures like those in Russia and China are changing. Social character is changing—but not because of a crisis like the current refugee crisis in Europe. It's changing because of modern technology and opportunities for a life that connects work, individuality and society. The future is coming; it's our task to set basics for the coming structural changes and support responsive handling of future technology.

References

Abdi, S. (2014): Self Driving Car—Google zeigt ersten Prototypen. Internet: http://www.computerbase.de/2014-05/self-driving-car-google-zeigt-ersten-prototypen. Access at 30.10.2015.

acc. (2016): Definition of Connected Car—What is the Connected Car. Internet: http://www.autoconnectedcar.com/definition-of-connected-car-what-is-the-connected-car-defined/. Access at 14.02.2016.

Ahmad, I. (2015): The Effects of Social Networking upon Society. Internet: https://www.linkedin.com/pulse/effects-social-networking-upon-society-idrees-ahmad. Access at 12.02.2016.

BDI (Ed.). (2015): *Die digitale Transformation der Industrie—Was sie bedeutet. Wer gewinnt. Was jetzt zu tun ist.* Berlin: Bundesverband Deutsche Industrie.

Bendel, O. (2016): Soziale Medien. Internet: http://wirtschaftslexikon.gabler.de/Definition/soziale-medien.html. Access at 12.02.2016.

Bourse. (2016): Tesla Motors Inc. Registered Shares. Internet: http://boersen.manager-magazin.de/mm/kurse_einzelkurs_uebersicht.htn?i=25382207. Access at 15.02.2016.

Bpb. (2010a): Soziale Netzwerke. Internet: http://www.bpb.de/nachschlagen/zahlen-und-fakten/globalisierung/52777/soziale-netzwerke. Access at 12.02.2016.

Bpb. (2010b): Musik. Internet: http://www.bpb.de/nachschlagen/zahlen-und-fakten/globalisierung/52783/musik. Access at 12.02.2016.

Bpb. (2011): Wie Medien genutzt werden und was sie bewirken. Internet: http://www.bpb.de/izpb/7543/wie-medien-genutzt-werden-und-was-sie-bewirken?p=all. Access at 15.02.2016.

Bundesregierung. (2016): Die Automobilindustrie—Eine Schlüsselindustrie unseres Landes. Internet: http://www.bundesregierung.de/Content/DE/Magazine/emags/economy/051/sp-2-die-automobilindustrie-eine-schluesselindustrie-unseres-landes.html. Access at 10.01.2016.

Damrow, S. (2012): Soziale Netzwerke wie Facebook—Bedeutung und Auswirkung. Internet: http://medienbildung.hypotheses.org/676. Access at 12.02.2016.

destatis. (2016): Arbeitsmarkt. Internet: https://www.destatis.de/DE/ZahlenFakten/Indikatoren/Konjunkturindikatoren/Arbeitsmarkt/arb210.html. Access at 15.02.2016.

Eckl-Dorna, W. (2016): 2016 wird das härteste Jahr, das Elon Musk je erlebt hat. Internet: http://www.manager-magazin.de/unternehmen/autoindustrie/tesla-zittern-vor-den-zahlen-elon-musk-unter-druck-a-1076741.html. Access at 15.02.2016.

Focus. (2012): Facebook brachte Proteste in Arabien auf die Straße. Internet: http://www.focus.de/digital/computer/arabischer-fruehling-facebook-brachte-proteste-in-arabien-auf-die-strasse_aid_746934.html. Access at 12.02.2016.

GF. (2015): Globalisierung durch Kommunikation und Internet. Internet: http://www.globalisierung-fakten.de/globalisierung-informationen/gruende/globalisierung-%20durch-kommunikation-und-internet/. Access at 29.10.2015.

GF. (2016a): Auswirkungen der Globalisierung. Internet: http://www.globalisierung-fakten.de/globalisierung-informationen/auswirkungen-der-globalisierung/. Access at 12.02.2016.

GF. (2016b): Definition Globalisierung. Internet: http://www.globalisierung-fakten.de/globalisierung-informationen/definition/. Access at 12.02.2016.

Gründerszene. (2016): Social-Media. Internet: http://www.gruenderszene.de/lexikon/begriffe/social-media. Access at 10.02.2016.

Handelsblatt. (2015): Made in Germany ist in Gefahr. Internet: http://www. handelsblatt.com/unternehmen/industrie/vw-abgasaffaere-made-in-germany-ist-in-gefahr/12353418.html. Access at 10.01.2016.

Hess, T. (2013): Digitalisierung. Internet: http://www.enzyklopaedie-der-wirtschaftsinformatik.de/lexikon/technologien-methoden/Informatik--Grundlagen/digitalisierung. Access at 14.02.2016.

Hotten, R. (2015): Volkswagen—The scandal explained. Internet: http://www.bbc. com/news/business-34324772. Access at 10.01.2016.

IKUD. (2009): Globalisierung—Definition. Internet: http://www.ikud.de/glossar/ globalisierung-definition.html. Access at 13.02.2016.

IuG. (2010): Netzwerke früher und heute. Internet: http://www.informatik.uni-oldenburg. de/~iug10/sn/html/content/netzwerke.html. Access at 12.02.2016.

Kirkpatrick, D. (2011): Der Facebook-Effekt. Internet: http://www.focus.de/digital/ internet/dld-2011/debate/tid-21059/soziale-netzwerke-der-facebook-effekt_ aid_592034.html. Access at 12.02.2016.

Lühmann, M. (2016): Vom Rand in die Mitte. Internet: https://www.freitag.de/ autoren/der-freitag/vom-rand-in-die-mitte. Access at 15.02.2016.

Pallenberg, S. (2015): Anstatt Apple Car—Apple baut iPads fuer alle Autos. Internet: http://www.mobilegeeks.de/artikel/apple-car-apple-auto-ipad/. Access at 14.02.2016.

pcmag. (2016): Definition of: connected car. Internet: http://www.pcmag.com/ encyclopedia/term/66377/connected-car. Access at 14.02.2016.

Scheuer, S. (2013): Wie China die sozialen Netzwerke auswertet. Internet: http://www. zeit.de/digital/datenschutz/2013-10/china-soziale-netzwerke-weibo-zensur. Access at 12.02.2016.

Schmidt, B. (2015): Apple plant eigenes Auto für 2019. Internet: http://www.autobild. de/artikel/apple-icar-titan-2019-vorschau-718909.html. Access at 14.02.2016.

Slavik, A., Süddeutsche Zeitung. (2013): Label ohne Wert. Internet: http://www. sueddeutsche.de/wirtschaft/streit-um-made-in-germany-label-ohne-wert-1. 1743839. Access at 10.01.2016.

Smirnova, J. (2014): Russland will Kreml-Kritiker im Internet kontrollieren. Internet: http://www.welt.de/politik/ausland/article127204379/Russland-will-Kreml-Kritiker-im-Internet-kontrollieren.html. Access at 12.02.2016.

Spengler, R. (2012): Soziale Netzwerke schädigen soziale Fähigkeiten. Internet: http:// www.welt.de/wirtschaft/karriere/leadership/article106568479/Soziale-Netzwerke-schaedigen-soziale-Faehigkeiten.html. Access at 12.02.2016.

Spiegel. (2015): Exportrekord—Deutschland knackt wieder die Billionenmarke. Internet: http://www.spiegel.de/wirtschaft/soziales/exporte-neuer-rekord-im-jahr-2014-a-1017416. html. Access at 10.01.2016.

Spiegel. (2016): Wegen Terrordrohungen—Twitter sperrt mehr als 125.000 Profile. Internet: http://www.spiegel.de/netzwelt/netzpolitik/is-terror-twitter-sperrt-mehr-als-125-000-profile-a-1075996.html. Access at 15.02.2016.

Statista. (2015a): Die 20 größten Exportländer weltweit im Jahr 2014 (in Milliarden US-Dollar). Internet: http://de.statista.com/statistik/daten/studie/37013/umfrage/ ranking-der-top-20-exportlaender-weltweit/. Access at 10.01.2016.

Statista. (2015b): Automobilindustrie in Deutschland. Internet: http://de.statista. com/statistik/faktenbuch/339/a/branche-industrie-markt/automobilindustrie/ autoindustrie/. Access at 10.01.2016.

Statista. (2015c): Google & Co. als Autobauer? Internet: http://de.statista.com/infografik/3806/tech-konzerne-als-autobauer/. Access at 10.01.2016.

Statista. (2016a): Statistiken und Studien zum Thema Tesla Motors. Internet: http://de.statista.com/themen/2418/tesla-motors/. Access at 15.02.2016.

Statista. (2016b): Aktuelle Statistiken zum Thema Soziale Online-Netzwerke. Internet: http://de.statista.com/themen/1842/soziale-netzwerke/. Access at 12.02.2016.

Statista. (2016c): Ranking der Länder mit höchster durchschnittlicher Nutzungsdauer von Social Networks weltweit im Jahr 2015. Internet: http://de.statista.com/statistik/daten/studie/160137/umfrage/verweildauer-auf-social-networks-pro-tag-nach-laendern/. Access at 12.02.2016.

Stocker, F. (2014): Verfall deutscher Tugenden gefährdet Wohlstand. Internet: http://www.welt.de/wirtschaft/article133049085/Verfall-deutscher-Tugenden-gefaehrdet-Wohlstand.html. Access at 10.01.2016.

Stöcker, C. (2016): Einfluss auf die Gesellschaft—Radikal dank Facebook. Internet: http://www.spiegel.de/netzwelt/netzpolitik/filterblase-radikalisierung-auf-facebook-a-1073450.html. Access at 12.02.2016.

Stumberger, R. (2012): Wie neue Medien den Informationsfluss und damit die Gesellschaft verändern. Internet: http://www.heise.de/tp/artikel/36/36100/1.html. Access at 10.01.2016.

University of Media Stuttgart. (2015): Die kulturelle und gesellschaftliche Globalisierung. Internet: https://www.hdm-stuttgart.de/mediensoziologie/globalisierung/kultur.htm. Access at 29.10.2015.

Welt. (2010): Chinesen umgehen die staatliche Internetzensur. Internet: http://www.welt.de/wirtschaft/webwelt/article5871314/Chinesen-umgehen-die-staatliche-Internetzensur.html. Access at 12.02.2016.

Welt. (2015a): Volkswagen wird zum globalen Mega-Arbeitgeber. Internet: http://www.welt.de/wirtschaft/article137317721/Volkswagen-wird-zum-globalen-Mega-Arbeitgeber.html. Access at 10.02.2016.

Welt. (2015b): Nahles erwartet deutlich steigende Arbeitslosenzahlen. Internet: http://www.welt.de/politik/deutschland/article147990057/Nahles-erwartet-deutlich-steigende-Arbeitslosenzahlen.html. Access at 15.02.2015.

Wied, A. (2016): Made in Germany—Historie einer Herkunftsbezeichnung. Internet: http://www.made-in-germany.biz/ueber-uns/made-in-germany.html. Access at 10.01.2016.

Wiesmüller, M. (2014): iPhone-Nutzung im Auto—Apple CarPlay im Test. Internet: http://www.computerbild.de/artikel/cb-News-Connected-Car-Apple-CarPlay-Test-11025402.html. Access at 14.02.2016.

3 The paradigm shift: From a static to a dynamic media business model

Zvezdan Vukanovic

3.1 Introduction: The core, constitutive and complementary semantics and concepts of the static and dynamic media business model comparison

Today, the transition from digital media business strategy to digital media business processes has become much more of a challenge, as media business processes are now mainly digitised and ICT enabled. Consequently, today's ICT-based media businesses' environments and management are more dynamic, characterised by ongoing fast changes and severe stakeholders' pressure. Therefore, the dynamic business model has risen to prominence as a conceptual and contextual tool of 'alignment' to fill the gap between corporate strategy and business processes, including their Web, Internet and digital infrastructure, providing crucial harmonisation between these organisational layers.

The author argues that the business model is an essential conceptual tool of alignment in digital business. More specifically, it represents an intermediate layer between business strategy and ICT-enabled business processes, filling the missing link created by the complex and digitised environment. Accordingly, a successful business should treat business strategy, business model, and business processes along with their IS, Information Systems as a harmonised package.

Broadly, two major and different conceptual types/approaches and applicative uses of business models can be identified. The first refers to a static approach. The second use of the concept represents a dynamic/transformational approach. The static view of a business model allows us to build typologies and study its relationship with performance (Demil and Lecocq 2010). The dynamic/transformational view deals with the major managerial question of how to change it (Demil and Lecocq 2010). The static model captures the target business and the key components of a business plan (Morris et al. 2005), while the dynamic/evolutionary model describes how a business evolves via four approaches: enhancing, extending, expanding and exiting (Applegate et al. 2003). Moreover, the transformational model addresses change and innovation in the organisation or in the model itself (Demil and Lecocq 2010). Static business models ignore both dynamics and change (Palo and Tähtinen 2011). Such models may not help companies to demonstrate their business in uncertain contexts (Doganova and Eyquem-Renault 2009). Likewise, dynamics—a modular component inherent in digital media–, ICT- and telecommunications-related products and services, is absent from the models. The author notes that ICT and media businesses must adopt digital media as the center of their business operations and incorporate their digital aspects such as efficient interoperability, better scalability and effective innovation into the main enterprise.

3.2 Methodology

The author uses a meta-analysis method combining and contrasting the data evidence and results from 80 relevant studies in the field of digital, media and ICT business models (including studies in *Long Range Planning, The Strategic Management Journal, The MIT Sloan Management Review, The Academy of Management Journal, The Harvard Business Review, The European Journal of Information Systems, The American Economic Review, The Journal of Business Research, The California Management Review* and *The Journal of Marketing*) to examine the key research questions, solutions and paradigm shifts found within the studies. This approach has several advantages:

- The precision and accuracy of the estimates can be improved as more data is used.
- The presence of publication bias can be investigated.
- Hypothesis testing can be applied to summary estimates.

3.3 The Benefits of a Dynamic Media Business Model

Dynamic media business models must be designed around ways of improving the customer experience, not around ways of improving the performance of the current business model. A dynamic media business model will help target benefits in three key areas:

- Higher profitability—by reducing operational costs through common platforms and integrated business processes, enabling the enterprise to leverage identifiable unique content and consumer experience assets across multiple platforms—better scalability—through digital workflows, rights and royalties solutions that can support millions of digital transactions and through the digital consolidation of physical format archives to reduce costs and boost commercial exploitation.
- More effective and continual innovation—through integration and automation, freeing up time for staff to collaborate and generate new ideas.

Repositioning digital as the engine of the business enables it to rebalance its skills and capabilities around control and data and to drive new services.

Furthermore, the move from the physical world to the digital world means more than simply replicating physical goods in digital forms or even creating new digital products. The move to digital also requires a major shift in a business's revenue model (Macnamara 2010).

In an era of hypercompetitive and volatile markets, a successful dynamic media business model disrupts not only channels, operations and products but established revenue models as well. The dynamic business model is a catalyst as well as a harbinger of the new digital mediascape. The dynamic business model should be regarded as an evolutionary/longitudinal/causal strategy process and dynamics. Accordingly, digital models are constantly evolving, as consumers, business technologies and customer preferences are changing rapidly (Macnamara 2010). One of the most contentious and pressing issues concerning the media in the early 21st century is identifying viable business models, with widespread reports that the 20th-century business models underpinning press, radio and television are collapsing because of 'audience fragmentation', driven by an ever-widening range of choice in media content and sources on the Internet (Macnamara 2010). Increasing demand for technological innovation in the media industry is driving paradigmatic changes in digital media business models (Macnamara 2010). Moreover, in today's increasingly global and competitive market, business models in the media and ICT industry are shifting from incremental to disruptive and transformational business model innovation (Macnamara 2010). Thus, the identification of sustainable media business models is an urgent priority, as the continuing decline in audiences and the collapse of media organisations pose a major threat to journalism and society, with scholars agreeing that the further erosion of quality journalism threatens democracy. Future media business models also have major implications for the advertising industry and a wide range of content producers (Macnamara 2010).

At this stage, no consensus or even widespread agreement has emerged on any alternative business model, and many of those proposed require further development and analysis. In that process, economic feasibility and market acceptance need to be balanced sensitively. However, the diversity of types of media content and media users' needs and preferences indicates that a 'one-size-fits-all approach' is unlikely to ensure media survival—or, better, reform and renewal. As Carr (2010) suggests, the best way forward may be a hybrid model involving diversification to create multiple revenue streams developed to suit each medium and its operations.

3.4 The impact of the exponential growth of ICT networks, traffic, and web/multimedia/hypermedia content on the formation of the dynamic media business model

Economic growth and technology are inextricably linked. Viewed longitudinally, technology is probably the most powerful influence on business models in the media sector, and the quality of an organisation's response to changes in this domain is probably one of the most important determinants of strategic outcomes. Also viewed longitudinally, the pattern of technology development for the sector is consistent: technology gives and technology takes away, but it seldom takes everything away. Technological innovations supplement, rather than replace, previous technologies. The previous medium is not destroyed but progressively undermined (Küng 2011). Additionally, businesses are, like media technologies in general, always already remediated: when new models emerge, old models are supplemented and only rarely displaced (Deuze 2011).

Current economic conditions are fostering investment in technology as emerging markets ramp up their demand for technology to fuel growth and advanced markets seek new ways to cut costs and drive innovation. This becomes a virtual circle, as digital technologies drive consumer income and demand, education and training and the efficient use of capital and resources—leading to increased economic growth, particularly in emerging markets. Executives must be aware of the new challenges facing their firms as market momentum accelerates. Moreover, the 2007 recession and financial crisis caused a seismic shift, reshaping the global business landscape and producing sluggish growth in the West, as well as a shift in power to the East, with value-driven customers and rising risks everywhere. At the same time, the downturn has accelerated the adoption of the cutting-edge technologies (IOT, cloud computing, broadband internet, smartphones and so on).

A leitmotif throughout this chapter is the fast-changing context of the media industry. While current global markets are subject to greater turbulence and complexity at higher velocities, the urgency to respond and adapt depends on media multinationals' tailored strategies of mass customisation, multimedia optimisation, downstream production and adaptive and innovative business models. Accordingly, media corporations should adapt to the fast, tectonic, unparalleled, unprecedented and seismic technological, market and demand developments, building a competitive as well as sustainable advantage because market dynamics make existing capabilities obsolete tomorrow (van Kranenburg and Ziggers 2013).

Instead of trying to create stability, media corporations must actively work to disrupt their own advantages and the advantages of competitors by continuously challenging existing capabilities. This involves the continuous search for improvement along a fixed production function, while the latter element also requires discontinuous shifts from one production function to another that is more profitable. Consequently, a media firm only incrementally adapts its existing business model, emphasising process efficiency and effectiveness. The challenge for media corporations is to develop and to incorporate new business models, such as the innovation-centered business model, to fulfil the new requirements and demands. This approach enables media corporations to be really innovative and to develop new capabilities and resources to sustain their competitive position (van Kranenburg and Ziggers 2013).

The accelerating growth of digital media and e-business has raised interest in transforming traditional business models via developing new ones that better exploit the opportunities enabled by disruptive technological innovations. One of the major impacts of e-business on traditional business practices has been the multiplication of possible business configurations (networked multi-platforms), which increases consumer choice, as well as the architectural implementation of business models and managerial decisions. Thus, the four key technologies (digital megatrends) that are bringing the new digital economy to maturity are mobility, cloud computing, business intelligence and social media.

3.5 The influence of ICT exponential development on the digital media business model

After the beginning of the commercial usage of the Internet, the average number of published academic articles increased by approximately 4000%. Businesses and consumers were expected to add approximately 40 exaflops of computing capacity in 2014, up from 5 in 2008 and less than 1 in 2005 (Dobbs et al. 2014). These extraordinary advances in capacity, power and speed are fueling the rise of artificial intelligence, reshaping global manufacturing (George et al. 2014) and turbocharging advances in connectivity. This will be further dynamised through the current and future development and application of the Internet of things or Industry 4.0—in which the physical world becomes a type of information system—through sensors and actuators embedded in physical objects and linked through wired and wireless networks via the Internet protocol. Moreover, advances in wireless networking technology and the greater standardisation of communications protocols make it possible to collect data from these sensors almost anywhere at any time.

3.6 Trends in Global IP Traffic Growth

The Cisco® Visual Networking Index (VNI) document *The Zettabyte Era—Trends and Analysis* presents some of the main findings of Cisco's global IP traffic forecast and explores the implications of IP traffic growth for service providers. Moreover, the document reveals that global IP traffic has increased fivefold over the past five years and will increase threefold over the next five years. Overall, IP traffic will grow at a compound annual growth rate (CAGR) of 23% from 2014 to 2019. Two-thirds of all IP traffic will originate from non-PC devices by 2019. In 2014, only 40% of total IP traffic originated from non-PC devices, but by 2019, the non-PC share of total IP traffic will grow to 67%. PC-originated traffic will grow at a CAGR of 9%, and TVs, tablets, smartphones and machine-to machine (M2M) modules will have traffic growth rates of 17%, 65%, 62% and 71%, respectively. In 2014, wired devices accounted for the majority of IP traffic, at 54%. Global

Internet traffic in 2019 will be equivalent to 66 times the volume of the entire global Internet in 2005. Globally, Internet traffic will reach 37 gigabytes (GB) per capita by 2019, up from 15.5 GB per capita in 2014.

The number of devices connected to IP networks will be more than three times the global population by 2019. There will be more than three networked devices per capita by 2019, up from nearly two networked devices per capita in 2014. Accelerated in part by the increase in devices and the capabilities of those devices, IP traffic per capita will reach 22 GB per capita by 2019, up from 8 GB per capita in 2014. Broadband speeds will more than double by 2019. By 2019, global fixed broadband speeds will reach 42.5 megabits per second (Mbps), up from 20.3 Mbps in 2014. Globally, IP video traffic will be 80% of all IP traffic (both business and consumer) by 2019, up from 67% in 2014. This percentage does not include the amount of video exchanged through peer-to-peer (P2P) file sharing. Internet video to TV grew 47% in 2014. This traffic will continue to grow at a rapid pace, increasing fourfold by 2019. Internet video to TV will be 17% of consumer Internet video traffic in 2019, up from 16% in 2014. Consumer video-on-demand (VoD) traffic will nearly double by 2019. The amount of VoD traffic in 2019 will be equivalent to 7 billion DVDs per month. Globally, mobile data traffic will increase 10-fold between 2014 and 2019. Mobile data traffic will grow at a CAGR of 57% between 2014 and 2019, reaching 24.3 exabytes per month by 2019. Global mobile data traffic will grow three times faster than fixed IP traffic from 2014 to 2019. Global mobile data traffic was 4% of total IP traffic in 2014 and will be 14% of total IP traffic by 2019.

IP traffic is growing fastest in the Middle East and Africa, followed by the Asia Pacific region and Central and Eastern Europe. Total Internet traffic has experienced dramatic growth in the past two decades. More than 20 years ago, in 1992, global Internet networks carried approximately 100 GB of traffic per day. In 2014, global Internet traffic reached 16,144 GBps. By 2019, it is projected that the global Internet traffic will amount to 51,794 GBps. Globally, IP traffic will reach 22 GB per capita by 2019, up from 8 GB per capita in 2014, and Internet traffic will reach 18 GB per capita by 2019, up from 6 GB per capita in 2014. Importantly, the global average broadband speed continues to grow and will more than double from 2014 to 2019, from 20.3 to 42.5 Mbps. Globally, the average mobile network connection speed in 2014 was 1.7 Mbps. The average speed will double and will be nearly 4 Mbps by 2019. Moreover, global Wi-Fi connection speeds originated from dual-mode mobile devices will nearly double by 2019. The average Wi-Fi network connection speed (10.6 Mbps in 2014) will exceed 18.5 Mbps in 2019. Globally, there will be nearly 341 million public Wi-Fi hotspots by 2018, up from 48 million hotspots in 2014, a sevenfold increase (iPass Inc. and Maravedis and Rethink Study 2014). Wi-Fi is also on the move, becoming available on 60% of planes and 11% of trains by 2018. This compares to only 16% of planes and 3% of trains equipped with Wi- Fi in 2014. Community 'homespot' public Wi-Fi hotspots will see the most explosive growth, rising from just under 40 million in 2014 to over 325 million in 2018. Accordingly, between 2000 and May 2016, the Web has grown 60-fold, from 17.08 million to 1.03 billion users.

3.7 The anticipation of global market shifts in real time

The industries most affected by digital transformation include IT (72%); telecommunications (66%); entertainment, media and publishing (65%); retail (48%); and banking (47%) (*The New Digital Economy* 2011). Itami and Nishino (2010) consider that a business model contains what the business does and how the business makes a profit. The business model describes conceptually corporate innovation, resources, the market and value. In addition, the business model may derive from the analysis of market opportunity, product and services, competitive dynamics or strategies (Applegate et al. 2003).

The overall essence, as well as the ultimate goal and objective of a firm's business model, is to exploit a business opportunity by creating value for its customers/stakeholders, enticing them to pay for that value and converting those payments into (Afuah and Tucci 2001; Applegate 2001; Huarng and Yu 2011; Petrovic et al. 2001; Teece 2010; Zott and Amit 2010;). A business model should reflect the financial conditions in a business (Dubosson-Torbay et al. 2002). In other words, a business model should translate the conceptual model into numbers (Meyer and Crane 2010). Thus, a financial model, consisting of cost, revenue and profit, serves as the second tier of the two-tier business model. To become sustainable, businesses may need to adapt their business models as time goes by (Dahan et al. 2010).

The digital revolution is the most challenging transformation shaking the traditional-conservative business models (analogue) and establishing new online/emerging networked multi-platforms. Digital design evolves into digital architecture, networks, architectural multi-platforms and consequently an ecosystem. From the analogue axiomatic principle/ground rule that content/distribution is king, the discussion is now evolving towards the view that choice/access/apps is king. Moreover, the digital media business paradigm is further emphasised via different media content consumption and distribution patterns; these include access vs. content, franchises over networked and multi-platform distribution channels, free vs. pay/premium, broadcast/printed journalism vs. drone journalism/online journalism/web journalism, user vs. prosumer, producer vs. produser, traditional social media networks (Facebook, Twitter, YouTube) vs. temporary social media networks (Snapchat, WhatsApp) and many others. Additionally, the Internet of things creates a more synergetic and convergent added-value network. More specifically, it personalises the business context and value exchange, creating more effective network effects between the potential prosumers and the applications/services/products on offer.

3.8 The dynamic media business model

Characterising the business model as dynamic (Hedman and Kalling 2003; MacInnes 2005) is essential mainly because many industries today, such as the media, ICT and telecommunications, are undergoing continuing revolutions driven by innovative technologies; globalisation, including deregulation; and market changes. Indeed, the business environment has been greatly transformed. Unlike the traditional world of business, which is characterised by stability and low levels of competition, the world of digital business is complex, granular, networked, modular and dynamic while also displaying high levels of uncertainty and competition. As a result, in the more complex and sometimes unique digital business environment, the business model needs to be both explicit and more flexible.

An evolving dynamic/networked/modular business model consists of strategic objectives, missions and structures (Hambrick and Fredrickson 2001; Porter 1996); target markets (scope and market segment) and the business value chain/network/proposition (alliances, partnerships, product/service offering) (Achrol and Kotler 1999; Anderson et al. 1994); key intra- and inter-organisational operational processes and resources (capabilities and assets) (Bartlett and Ghoshal 1995; Barney 1991; Nelson and Winter 1982); a finance and accounting system; and a cost and revenue model (cost and revenue streams, pricing policy) (Norton and Kaplan 1992). Moreover, in Table 3.1, the author outlines the key building blocks of dynamic and static business models.

The advent of the Internet represents a crucial landmark in the evolution of digital media. Accordingly, the digital (new) media global landscapes landscapes demand new business model maps. Media companies are trying to face up to the challenges of this emerging scenario, as new consumers and new markets are transforming traditional business models into dynamic media business markets. As a consequence, media corporate players are moving strategically, and the whole audiovisual product value chain will be readjusted (Guerrero et al. 2013). The influence of this multi-platform audiovisual model on production is inextricably bound up with the question of the business models that may enable a recouping of the costs involved. A key aspect of the design of any business model is the identification of revenue streams. These network-based, symbiotic, market-driven relationships among the various entities and social media are the genesis of sustainable business models for the emerging social media industry. The dynamic business model is thus actively co-created between the various actors/platforms involved.

3.9 The main paradigm shifts in new/social media versus old/traditional media

Although both the old/traditional and the new/social media can reach small or large audiences, there are many fundamental differences in terms of the competitive advantage in distribution, production, technology and market

Table 3.1 The common denominators of the major paradigmatic shifts from the static to the dynamic/transformational media business model building blocks

Static Media Business Model Building Blocks	Dynamic/Transformational/Networked/Modular Media Business Model Building Blocks
Analogue media	Digital media
Organisational design	Organisational architecture evolving into organisational ecosystems and smart grid networks
One-sided market	Two-sided/multi-platform/network/market
Upstream supply chain (push marketing, low-cost producers)	Downstream supply chain (customisation, targetisation, high margins)
Top-down content production/ distribution	Bottom-up content production/distribution
One-to-many content distribution	Many-to-many content distribution
Linear, one-way communication	Interactive communication evolving into immersive communication
Reaching the audience	Connecting the audience
Passive users/consumers	Active users—produsers and prosumers
Mass audience	Audience fragmentation or disaggregation
Less available and accessible consumption to the public	On-demand access
Low level of collaborative content sharing	High level of collaborative content sharing—user-generated content (UGC)
Bundling	Content; P2P; tagging; folksonomy; Big Data analytics; IOT (Internet of things); social networks; WOT (Web of things); wearable technology; locative/mobile media complementarities; and vendor lock-in strategies
Broadcasting	Broadband, narrowcasting, microcasting and egocasting
Content and distribution are king	Choice, share, access and application are king
Competition	Co-opetition
Freemium	Premium
Industrial, tangible economy— economies of scale	Information, intangible economy—economies of scope, long-tail economics, digital economics, network economics, information economics, experience economics
Push market revenue model	Pull market revenue model—'behavioural targeting', 'advergaming', 'gamification', 'product placement', micro-payment, paywall, content repurposing, sale of data and 'asynchronous ads'
Two-dimensional media	3D media
Web 1.0 and Web 2.0	Web 3.0 (semantic web) and Web 4.0 (symbiotic web)
Symmetric information flow	Asymmetric information flow
First build a marketplace, then a community	First build a community, then a marketplace
Attention span is longer	Attention span is shorter
Owning the accessed content	Sharing the accessed content
Searching the data	Searching the metadata
Hardware-based media	Software-based (cloud-based) media
Demand is king	Choice is king
Connect individual with the information/content/product	Share content and experience among groups
Information-based service	Conversation/communication-based service
Place-bounded media	Space-bounded media
Individual/one-screen media	Multi-screen media
Value is contained in the transaction	Value is contained in the relationship
Usage-based pricing	Access-based pricing

targeting that favour new/social media over old/traditional media. In Table 3.1, these distinguishing differences are exposed in order to more effectively outline the major conceptual differences between new and old media.

3.10 The positioning modeling principles of the dynamic media business model

The digital era has meant that the availability of appropriate levels of information and knowledge has become critical to the success of any business. Organisations need to adapt in order to survive and succeed as their business domains, processes and technologies change in a world of increasing environmental complexity. Enhancing their competitive positions by improving their ability to respond quickly to rapid environmental changes with high-quality business decisions can be supported by adopting suitable business models for this new world of digital business. However, the main reason behind this confusion is the shift that the business world has experienced from the traditional ways of doing business to the new ways of digital business, which feature a high level of complexity and rapid change. This transformation has created a gap between strategy and processes that calls for new ways of thinking about business models.

The modern media-, telecommunications-, and ICT-based world of business imposes a vital need for business models with high levels of adaptability to accommodate ongoing changes more efficiently. Within today's business environment, the business model should also enjoy dynamicity to cope successfully with the continuous changes it faces. The granularity and modularity of the dynamic business model imply flexibility in its related functions such as design, architecture, management, evaluation and change and also facilitate the reusability of the components for new business models. This highlights the concept as an efficient and effective framework essential to digital organisations. This subfield of research is still unexplored; therefore, a theoretical as well as a practical investigation and delineation of this particular area would be very useful.

3.11 The evolution and position of the media business model within the new world of digital business

Media business model researchers are attempting to determine its meaning, boundaries, components and relationships with other business aspects, such as business processes and business strategy. There is already some consensus regarding the differences between the business model and the process model (Pateli and Giaglis 2003; Morris et al. 2005). However, the debate on the difference between the business model and the business strategy has not been resolved (Porter 2001; Stähler 2002; Pateli and Giaglis 2004). Some researchers see them as identical and use the terms interchangeably: for example, Kallio et al. (2006) depict business model components as a set of business strategies. Other researchers suggest that even though both concepts are related, they represent different levels of information, useful for different purposes. They see the business model as an interface or an intermediate theoretical layer between the business strategy and business processes (Osterwalder 2004; Tikkanen et al. 2005; Rajala and Westerlund 2005; Morris et al. 2005). Magretta (2002) argues that the business strategy explains how business organisations hope to do better than their rivals, while the business model describes how the pieces of a business all fit together.

The main reason behind this confusion is the shift that the business world has experienced away from the traditional forms of doing business to the new mechanisms of digital business (e-business), which has brought to the fore a high level of complexity and rapid change. This new world of digital business has created a gap between the business strategy and business processes. In this context, translating business strategy into business process has become much more of a challenge. Accordingly, the business model has risen to prominence as a conceptual tool of alignment to fill the gap that has been created in this world of digital business.

The business model facilitates the fit and an interface or an intermediate layer between business strategy and business processes. Furthermore, the business model enhances digital business managers' control over their business and enables them to compete better, due to the appropriate and necessary level of information that the business model provides. This level of information also extends digital business managers' knowledge of how a given business organisation will adapt its strategy, business model and business processes to cope with the complex, uncertain and rapidly changing digitalised environment. Thus, there are improvements in the organisation's abilities to achieve its strategic goals and objectives. This is because the information that the business model offers is neither highly aggregated, which it is in the case of the business strategy, nor highly detailed, which it is in the case of the operational business process model. The business model is by no means independent; it intersects with the business strategy as well as business processes.

3.12 Digital media business model perspectives: From place to space

Before the Internet, business operated primarily in a physical world of 'place': it was a world that was tangible, product-based and oriented toward customer transactions. Today, many industries—all moving at different rates—are shifting toward a digital world of 'space': it is a world that is more intangible, more service-based and application-based and oriented toward customer experience. In the world of 'space', the components of content, packaging and infrastructure have morphed (e.g., converged) and split (e.g., diverged). Content has mushroomed and is no longer strictly proprietary. The packaging has transformed into a consistent digital customer experience across many different devices. Infrastructure has morphed into a powerful combination of internal and external digital platforms—some of which are controlled by media content producers and some not.

The concept of a digital business model draws on previous research on business models, much of which focused on e-business (for example, Dubosson-Torbay et al. 2002; Mahadevan 2000; Gordijn et al. 2005; Gordijn and Akkermans 2001; Hedman and Kalling 2002; Menasce 2000; Swatman et al. 2006; Gordijn 2004; Osterwalder and Pigneur 2002; Shin and Park 2009; Chen and Ching 2002; Pigneur 2000; Currie 2004a, 2004b; Gordijn et al. 2000; Faber et al. 2003; Damanpour and Damanpour 2001; Argoneto and Renna 2010; Gordijn 2003; Jarvenpaa and Tiller 1999; Papakiriakopoulos et al. 2001; Lambert 2006a, 2006b; Pateli and Giaglis 2005). These e-business models can be regarded as a subset of business models (Vermolen 2010). In a highly globalised and competitive market, media and ICT enterprises need to strengthen their digital business model. However, digital business models can crash quickly because switching costs in the digital world are often lower than in the physical world.

3.13 The main components of a digital media business model

A digital media business model has three components: content (what is consumed?), customer experience (how is it packaged?) and platform (how is it delivered?). These three components work together to create a compelling customer-value proposition. Digital content includes digital products (e.g., software, movies and e-books), as well as information about price and use details and so on. The customer experience embodies what it is like to be a digital customer of the organisation. The platform consists of a coherent set of digitised business process, data, and infrastructure. The platform has internal and external components and may both deliver digital content to the customer and manage physical product delivery to the customer. Amazon's internal platforms include customer data and all the business processes that do not affect the customer, such as customer analytics, human resources, finance and merchandising. External platforms include the phones, tablets or computers that consumers use to research and purchase products, along with telecommunications networks and Amazon's partnerships with delivery companies like UPS that deliver physical products and generate text messages on delivery; all of these external platforms neatly integrate with Amazon's internal platforms. To achieve economies of scale with digital business models requires the development and reuse of digitised platforms across the enterprise (Weill and Ross 2009). Without such shared platforms, the IT units in companies implement a new solution in response to every business need, creating a spaghetti-like arrangement of systems that do meet specific customer needs but that are expensive and fragile—and that cannot scale enterprise-wide. Worse still, the customer experience suffers, as the customer gets a fragmented, product-based experience rather than a unified, multi-product experience.

3.14 Measuring the effectiveness of digital media business models' content, experience and platform

To better understand digital business models by industry, the author surveyed companies to assess the effectiveness of their content, experience and platform. For each of the three aspects of the digital business model (content, experience and platform), the author aggregated the answers to eight or nine survey questions to get a broad base for assessing effectiveness. The industry with the strongest effectiveness scores overall was IT software and services, while energy and mining and healthcare were among the poorest. Interestingly, the top financial performers in each industry also had better digital business model effectiveness. For example, in the financial services industry, companies in the top third of financial performers had 29%, 35% and 26% better content, experience and platform scores, respectively, than those in the bottom third.

3.15 The e-business model

The e-business sector is experiencing an unprecedented paradigm shift in terms of not being able to fully predict corporate sales, customer interaction, value-added network and so on. Therefore, corporations have to put a tremendous emphasis on quick response instead of traditional planning. The e-business model, as opposed to the old industrial model, is marked by fundamental rather than incremental change. Thus, it is impossible to plan an e-business model for the long term; instead, e-businesses must shift to a more flexible, predefined and anticipative model of planning (Malhotra 2000).

3.16 The research dynamics of transformational business models

The majority of the research on dynamic business models has been concerned with e-business and e-commerce, and there have been some attempts to develop convenient classification schemas. For example, definitions, components and classifications in e-business models have been suggested (Alt and Zimmermann 2001; Afuah and Tucci 2003). Researchers have also looked at the business model concept in the context of different domains. Accordingly, Linder and Cantrell (2000) and Magretta (2002) have applied the business model concept in the domains of business management and strategy (Bouwman et al. 2008; Al-Debei and Fitzgerald 2010); software; the telecom sector, including mobile technology along with its services industry (Rajala and Westerlund 2007); and e-government (Janssen et al. 2008).

Weill and Woerner (2013 : 29) define three converging trends in raising the stakes for the effectiveness of the enterprise's digital business model.

> The first is the continued march toward the digitization of ever-increasing aspects of business—incorporating more of your customers' experience,

executing more of your business processes and working together with partners in your value chain. The second trend is the increasing number of 'digital natives'—young current and future customers and employees—who expect a brilliant digital experience in all of their interactions with companies. The third trend is the dawning of the age of the customer voice, in which customers have a much stronger impact on enterprises via ratings of their services (such as the customer rating stars on Amazon and customer experience surveys) through Twitter and other social media comments.

In addition, Weill and Vitale (2001) define eight finite e-business models (direct customer, full-service provider, intermediary, whole enterprise, shared infrastructure, virtual community, value net integrator and content provider) based on a systematic and practical analysis of several case studies. As the business models move towards maturity, corporate and academic interest is shifting to the investigation of opportunities for more effective and efficient market exploitation of innovative and specifically topical business models. However, there is an alarming lack of empirical strategic models in the literature to structure, categorise and systematically codify knowledge in the area.

3.17 Conclusion

This chapter draws on an extensive review of the literature to propose the incremental/gradual evolution of business models from a static to a dynamic/networked/modular architecture framework.

The literature on business models recognises their applicative and market importance and influence on business environment dynamics and corporate strategy. Thus, the difference between the success and failure of transformative activities boils down to the firm's ability to change its business model effectively and in rhythm with the dynamics of the external business environment (Burgelman 1994; Siggelkow 2001). Moreover, there is a lack of studies that focus on the competition preceding digital business model changes. While many researchers have concentrated on conceptualising various generic components of the business model concept (Morris et al. 2005; Siggelkow 2001; Amit and Zott 2001), managers' conceptualisations of business models and their links to paradigmatic business model evolution have mostly escaped researchers' attention so far. Academic research (for a review, see, e.g., Tikkanen et al. 2005), in turn, has referred to business models particularly when dealing with the novel and systemic mechanisms and architectures through which business will be done vis-à-vis the greater business environment and industry networks (Zott and Amit 2008; Chesbrough and Rosenbloom 2002).

In an increasingly global and mobile digital media landscape, it is easier than ever to reach a large audience, but it is harder than ever to effectively connect with it. Thus, where old media's traditional preoccupation was to reach their audience, in the age of digital media globalisation, digital media companies have the twofold task of both reaching and connecting with the audience.

In summary, in the second decade of the 21st century, digital media are apparently becoming increasingly interactive, mobile, immersive and ubiquitous. Furthermore, the future of the media appears to be specifically oriented towards the establishment of networked, 3D, on-demand broadband and unicast, as well as multimedia and hypermedia models of distribution, communication and media—with content repurposing creation. Therefore, it is crucial that profitable digital media companies realise that, in a period of media divergence, they can successfully perform as vendors that lock in a top-down corporate process and a bottom-up consumer-driven process.

The digital business models influenced by digital convergence/divergence will focus on aggregate multi-platform distribution, digital platforms, complementarities, vendor lock-in, interoperable and networked media and ICT ecosystem, massive personalisation/customisation and user interfaces.

References

Achrol, R. S., & Kotler, P. (1999). Marketing in the network economy. *Journal of Marketing, 63*, 146–163.

Afuah, A., & Tucci, C. L. (2001). *Internet business models and strategies: Text and cases* (4th ed.). New York: Irwin/McGraw-Hill.

Afuah, A., & Tucci, C. (2003). *Internet business models and strategies* (2nd ed.). New York: McGraw-Hill.

Al-Debei, M. M., & Fitzgerald, G. (2010). The design and engineering of mobile data services: Developing an ontology based on business model thinking. In Pries-Heje, J., Venable, J. J., Bunker, D., Russo, N., DeGross, J. (Eds.), *Human benefits through the diffusion of information systems design science research*. Boston: Springer.

Alt, R., & Zimmermann, H. (2001). Introduction to special section—Business models. *Electronic Markets, 11*(1), 3–9.

Amit, R., & Zott, C. (2001). Value creation in e-business. *Strategic Management Journal, 22*(6–7), 493–520.

Anderson, J. C., Håkansson, H., & Johanson, J. (1994). Dyadic business relationships within a business network context. *Journal of Marketing, 58*(4), 1–15.

Applegate, L. M. (2001). *Emerging e-business models: Lessons from the field*. Boston: Harvard Business School.

Applegate, L. M., Austin, R. D., & McFarlan, F. W. (2003). *Corporate information strategy and management* (6th ed.). New York: McGraw Hill.

Argoneto, P., & Renna, P. (2010). Production planning, negotiation and coalition integration: A new tool for an innovative e-business model. *Robotics and Computer-Integrated Manufacturing, 26*(1), 1–12.

Barney, J. (1991). Firm resources and sustained competitive advantage. *Journal of Management, 17*(1), 99–120.

Bartlett, C. A., & Ghoshal, S. (1995). Changing the role of top management: Beyond systems to people. *Long Range Planning, 28*(4), 126.

Bouwman, H., De Vos, H., & Haaker, T. (2008). *Mobile service innovation and business models*. Berlin: Springer.

Burgelman, R. A. (1994). Fading memories: A process theory of strategic business exit in dynamic environments. *Administrative Science Quarterly, 39*, 24–56.

Carr, D. (2010, April 19). Government funding cannot save journalism. *The Nation*. www.thenation.com/doc/20100419/carr_video.

Chen, J. S., & Ching, R. K. (2002). A proposed framework for transitioning to an e-business model. *Quarterly Journal of Electronic Commerce, 3*(4), 375–389.

Chesbrough, H., & Rosenbloom, R. S. (2002). The role of the business model in capturing value from innovation: Evidence from Xerox Corporation's technology spin-off companies. *Industrial and Corporate Change, 11*(3), 529–555.

Currie, W. L. (2004a). Value creation from the application service provider e-business model: The experience of four firms. *Journal of Enterprise Information Management, 17*(2), 117–130.

Currie, W. (2004b). *Value creation from e-business models*. Oxford: Butterworth-Heinemann.

Dahan, N. M., Doh, J. P., Oetzel, J., & Yaziji, M. (2010). Corporate-NGO collaboration: Co-creating new business models for developing markets. *Long Range Planning, 43*(2), 326–342.

Damanpour, F., & Damanpour, J. A. (2001). E-business e-commerce evolution: Perspective and strategy. *Managerial Finance, 27*(7), 16–33.

Demil, B., & Lecocq, X. (2010). Business model evolution: In search of dynamic consistency. *Long Range Planning, 43*, 227–246.

Deuze, M. (2011). *Managing media work*. Thousand Oaks, CA: Sage.

Dobbs, R., Ramaswamy, S., Stephenson, E., & Viguerie, S. P. (2014, September). Management intuition for the next 50 years. *McKinsey Quarterly*.

Doganova, L., & Eyquem-Renault, M. (2009). What do business models do? Innovation devices in technology entrepreneurship. *Research Policy, 38*, 1559–1570.

Dubosson-Torbay, M., Osterwalder, A., & Pigneur, Y. (2002). E- business model design, classification, and measurements. *Thunderbird International Business Review, 44*(1), 5–23.

Faber, E., Ballon, P., Bouwman, H., Haaker, T., Rietkerk, O., & Steen, M. (2003). Designing business models for mobile ICT services. Paper presented at the 16th Bled Electronic Commerce Conference eTransformation, Bled, Slovenia.

George, K., Ramaswamy, S., & Rassey, L. (2014, January). Next-shoring: A CEO's guide. *McKinsey Quarterly*.

Gordijn, J. (2003). Why visualization of e-business models matters. Paper presented at the 16th Bled Electronic Commerce Conference eTransformation. Bled, Slovenia.

Gordijn, J. (2004). e-Business value modelling using the e3-value ontology. In W. Currie (Ed.), *Value creation from e-Business models* (pp. 98–128). Oxford: Elsevier Butterworth-Heinemann.

Gordijn, J., & Akkermans, H. (2001). Designing and evaluating eBusiness models. *IEEE Intelligent Systems, 16*(4), 11–17.

Gordijn, J., Akkermans, H., & Vliet, H. V. (2000). What's in an electronic business model? Paper presented at the 12th International Conference on Knowledge Engineering and Knowledge Management, Juan-les-Pins, France.

Gordijn, J., Osterwalder, A., & Pigneur, Y. (2005). Comparing two business model ontologies for designing e-business models and value constellations. Proceedings of 18th Bled Electronic Conference, Bled, Slovenia, Maribor, Slovenia.

Guerrero, E., Diego, P., & Pardo, A. (2013). Distributing audiovisual contents in the new digital scenario: Multiplatform strategies of the main Spanish TV networks. In M. Friedrichsen & W. Mühl-Benninghaus (Eds.), *Handbook of social media management: Value chain and business models in changing media markets*. Berlin: Springer.

Hambrick, D. C., & Fredrickson, J. W. (2001). Are you sure you have a strategy? *Academy of Management Executive, 15*(4), 48–59.

Hedman, J., & Kalling, T. (2002). *IT and business models: Concepts and theories*. Malmo: Liber Ekonomi.

Hedman, J., & Kalling, T. (2003). The business model concept: Theoretical underpinnings and empirical illustrations. *European Journal of Information Systems, 12*(1), 49–59.

Huarng, K. H., & Yu, T. H. K. (2011). Entrepreneurship, process innovation and value creation by a non-profit SME. *Management Decision, 49*(2), 284–296.

iPass Inc. and Maravedis and Rethink Study, 2014 https://www.ipass.com/blog/unique-ipass-study-reveals-global-wi-fi-explosion/

Itami, H., & Nishino, K. (2010). Killing two birds with one stone: Profit for now and learning for the future. *Long Range Planning, 43*, 364–369.

Janssen, M., Kuk, G., & Wagenaar, R. W. (2008). A survey of web-based business models for e-government in the Netherlands. *Government Information Quarterly, 25*(2), 202–220.

Jarvenpaa, S. L., & Tiller, E. H. (1999). Integrating market, technology, and policy opportunities in e-business strategy. *Journal of Strategic Information Systems, 8*(3), 235–249.

Kallio, J., Tinnilä, M., & Tseng, A. (2006). An international comparison of operator-driven business models. *Business Process Management Journal, 12*(3), 281–298.

Küng, L. (2011). Managing strategy and maximizing innovation in media organisations. In M. Deuze (Ed.), *Managing media work* (pp. 43–56). Thousand Oaks, CA: Sage.

Lambert, S. (2006a). A business model research schema. In *BLED 2006 Proceedings*, 43.

Lambert, S. (2006b). *Do we need a 'real' taxonomy of E-business models?* (School of Commerce Research Paper Series: 06-6). Adelaide, Australia: School of Commerce, Flinders University.

Linder, J., & Cantrell, S. (2000). *Changing business models: Surveying the landscape* (Working Paper). x: Accenture Institute for Strategic Change.

MacInnes, I. (2005). Dynamic business model framework for emerging technologies. *International Journal of Service Technology and Management, 6*(1), 3–19.

Macnamara, J. (2010). Remodelling media: The urgent search for new media business models. *Media International Australia, 137*, 20–35.

Magretta, J. (2002). Why business models matter. *Harvard Business Review, 80*(5), 86–92.

Mahadevan, B. (2000). Business models for internet-based e-commerce: An anatomy. *California Management Review, 42*(4), 55–69.

Malhotra, Y. (2000). Knowledge management for e-business performance: Advancing information strategy to 'Internet time'. *Information Strategy: The Executive's Journal, 16*(4), 5–16.

Menasce, D. (2000). Scaling for e-business. In *Proceedings of the 8th International Symposium on Modeling, Analysis and Simulation of Computer and Telecommunication Systems* (pp. 511–513). x: IEEE.

Meyer, M. H., & Crane, F. G. (2010). *Entrepreneurship: An innovator's guide to startups and corporate ventures.* Thousand Oaks, CA: Sage.

Morris, M., Schindehutte, M., & Allen, J. (2005). The entrepreneur's business model: Toward a unified perspective. *Journal of Business Research, 58*(6), 726–735.

Nelson, R. R., & Winter, S. G. (1982). The Schumpeterian tradeoff revisited. *American Economic Review, 72*, 114–132.

The new digital economy: How it will transform business: A white research paper produced in collaboration with AT&T, Cisco, Citi, PwC & SAP. (2011). Oxford: Oxford Economics.

Norton, D., & Kaplan, R. (1992). The balanced scorecard: Measures that drive performance. *Harvard Business Review, 70*(1), 52–58.

Osterwalder, A. (2004). *The business model ontology: A proposition in a design science approach.* Doctoral thesis, Présentée _a l'Ecole des Hautes Etudes Commerciales de l'Université de Lausanne.

Osterwalder, A., & Pigneur, Y. (2002, June 17–19). An e-business model ontology for modeling e-business. In C. Loebbecke, R. T. Wigard, J. Gricar, A. Pucihar & G. Lenart (Eds.), *Proceedings of the 15th Bled Electronic Commerce Conference—eReality: Constructing the eEconomy* (pp. 75–91). Bled, Slovenia: x.

Palo, T., & Tähtinen, J. (2011). A network perspective on business models for emerging technology-based services. *Journal of Business and Industrial Marketing, 26*(5), 377–388.

Papakiriakopoulos, D., Poylumenakou, A. K., & Doukidis, G. J. (2001). Building e-business models: An analytical framework and development guidelines. In *Proceedings of the 14th Bled Electronic Commerce Conference* (vol. 25, p. 26). Bled, Slovenia: x.

Pateli, A. G., & Giaglis, G. M. (2003). A framework for understanding and analyzing ebusiness models. In *Proceedings of 16th Bled eCommerce Conference on eTransformation* (pp. 329–348). Bled, Slovenia: x.

Pateli, A. G., & Giaglis, G. M. (2004). A research framework for analyzing eBusiness models. *European Journal of Information Systems, 13*(4), 302–314.

Pateli, A. G., & Giaglis, G. M. (2005). Technology innovation-induced business model change: A contingency approach. *Journal of Organizational Change Management, 18*(2), 167–183.

Petrovic, O., Kittl, C., & Teksten, D. (2001, 31 October–4 November). Developing business models for eBusiness. Paper presented at the International Conference on Electronic Commerce, Vienna.

Pigneur, Y. (2000). *The e-business model handbook.* Lausanne: École des HEC–Université de Lausanne.

Porter, M. (1996). What is strategy? *Harvard Business Review, 74*(6), 61–78.

Porter, M. E. (2001). Strategy and the internet. *Harvard Business Review, 79*(3), 62–79.

Rajala, R., & Westerlund, M. (2005). Business models: A new perspective on knowledge-intensive services in the software industry. In *BLED 2005 Proceedings*, 10.

Rajala, R., & Westerlund, M. (2007). Business models—A new perspective on firms' assets and capabilities: Observations from the Finnish software industry. *International Journal of Entrepreneurship and Innovation, 8*(2), 115–126.

Shin, J., & Park, Y. (2009). On the creation and evaluation of e-business model variants: The case of auction. *Industrial Marketing Management, 38*(3), 324–337.

Siggelkow, N. (2001). Change in the presence of fit: The rise, the fall, and the renaissance of Liz Claiborne. *Academy of Management Journal, 44*(4), 838–857.

Stähler, P. (2002). Business models as a unit of analysis for strategizing. In *Proceedings of the 1st International Workshop on Business Models*. Lausanne: x.

Swatman, P. M., Krueger, C., & van der Beek, K. (2006). The changing digital content landscape: An evaluation of e-business model development in European online news and music. *Internet Research, 16*(1), 53–80.

Teece, D. (2010). Business model, business strategy, and innovation. *Long Range Planning, 43*(2–3), 172–194.

Tikkanen, H., Lamberg, J. A., Parvinen, P., & Kallunki, J. P. (2005). Managerial cognition, action and the business model of the firm. *Management Decision, 43*, 789–809.

van Kranenburg, H., & Ziggers, G. W. (2013). How media companies should create value: Innovation centered business models and dynamic capabilities. In M. Friedrichsen & W. Mühl-Benninghaus (Eds.), *Handbook of social media management: Value chain and business models in changing media markets* (pp. 239–267). Berlin: Springer.

Vermolen, R. (2010). Reflecting on IS business model research: Current gaps and future directions. In *Proceedings of the 13th Twente Student Conference on IT* (pp. 291–299). Enschede, Netherlands: University of Twente.

Weill, P., & Ross, J. W. (2009). *IT savvy: What top executives must know to go from pain to gain*. Boston: Harvard Business School Press.

Weill, P., & Vitale, M. R. (2001). *Place to space: Migrating to eBusiness models*. Boston: Harvard Business School Press.

Weill, P., & Woerner, S. L. (2013). Optimizing your digital business model. *MIT Sloan Management Review, 54*(3), 71–78.

Zott, C., & Amit, R. (2008). The fit between product market strategy and business model: Implications for firm performance. *Strategic Management Journal, 29*(1), 1–26.

Zott, C., & Amit, R. (2010). Business model design: An activity system perspective. *Long Range Planning, 43*(2–3), 216–226.

4 Ten global trends: A literature review on the future of IT, media and the cultural industries

José María Álvarez-Monzoncillo,
Guillermo de Haro-Rodríguez, and
Javier López-Villanueva

4.1 Introduction

During the last decade, the media and cultural sectors have confronted many changes caused by digital convergence. The transformation of printed and analogue audio and video material into binary files has made it possible for different means of transmitting information to be digitally stored in the same devices. Moreover, they can now be distributed—legally or illegally—through the same channels. This has been encouraged by compression and recommendation algorithms and underpinned by the improvement of different storage and reproduction products, thanks to network effects and an even faster network. In parallel, global trends that are changing the boundaries of industries have emerged.

This chapter will analyse 10 global trends in the fields of IT, media and the cultural industries. We believe it is a good time to take stock of key research themes and summarise empirical findings. This should allow us to identify research gaps and future research directions.

In order to select these trends, we have followed the work of several consultancies and institutions (such as Gartner, which identifies the Top 10 Strategic Technology Trends for each year). This analysis was supplemented with searches in the ISI Web of Knowledge using as keywords the trends. The present literature review encompasses contributions from many fields, such as strategy, organisation theory, marketing, cultural economics and sociology.

Nevertheless, it is not easy to find a literature review on the researchers and studies that support these trends. The present research is therefore both timely and necessary. To cover that gap, we have selected the 10 most relevant trends and the most important authors talking about them to better understand their evolution. Their complementarities will illustrate the multiple dimensions and explanatory factors of our world.

4.2 Empowerment and lateral power

Much has been written about power across all disciplines. Aristotle considered it to be an ingredient of happiness. With the arrival of means of communication, the concept was permanently transformed. Yet, with their appearance, technology is not the only thing that has changed the way that power is wielded and lost. We often think that information and communication technologies (ICT) or economic changes are behind this trend, but sometimes shifts in expectations and values are just as important. Joseph Nye (2011) tells us about the transition from 'hard power' to 'soft power' in culture. There are three categories of 'soft power': culture, political values and foreign policies. Antonio Gramsci considered cultural hegemony to be a means of reinforcing the power of a nation state and of capitalism itself, whilst Michael Foucault associates power with surveillance.

'Power is the ability to direct or impede the current or future actions of other groups and individuals...it is what we exercise over others that leads them to behave in ways they would not otherwise have behaved' (Naím, 2013: 38). This definition, inspired by Robert Dahl, encompasses different ways of imposing one's will, such as influence, leadership, persuasion and coercion. As Ulrich Beck might say, we live a world that must manage fears.

At any rate, power is being eroded because it is becoming more and more difficult to maintain and because people have tools such as social networks and the Internet that weaken it. Yet the opposite might also be observed: the Internet empowers the people. However, there are additional factors that underpin the dilution of power in all societies, thus leading to the appearance of societies with greater freedom and opportunities for their citizens. There are three reasons behind this process according to Naim (2013: 32):

> the revolution of *more*, that is characterised by the increase in and abundance of everything...; the revolution of *mobility*, that refers to the fact that not only is there more of everything but "more" of everything (people, products, technology, money) is on the move; this is increasingly taking place and largely because it can be done at a lower cost and with a global reach, even to those locations that until recently were inaccessible; and the revolution of *mentality*, that reflects major changes in the way of thinking, expectations and aspirations that have accompanied these transformations.

Social change is being caused by social, cultural and economic factors. Jeremy Rifkin asserts that the traditional hierarchical organisation of economic and political power is making way for lateral power that is organised nodally in the Third Industrial Revolution. He states that 'social networks have bloomed ... creating a new distributed and collaborative space for sharing knowledge and spurring creativity and innovation across every field' (Rifkin, 2011: 165-166). Change is under way: 'Societies evolve and change by deconstructing their institutions under the pressure of new power relationships and constructing new sets of institutions that allow people to live side by side without self-destroying, in spite of their contradictory interests and values' (Castells, 2007: 258).

Max Weber was right when he insisted that bureaucratic power is based on the power of compliance with rules and regulations. On the Internet, this is the first thing that fascinates: how easy it is to bypass rules and/or create parallel systems of rules and regulations (netiquette). Parental orders, intellectual property, anonymous insults and the like are the trolls. Members of the so-called Internet generation have rules that are very different from those of past generations and are making appeals to change the world by calling for freedom, customisation, scrutiny, integrity, collaboration, entertainment, speed and innovation (Tapscott, 2009: 74).

The Internet fosters debates between contents generated by companies that must be commercialised in order to attempt to generate profits and non-profit contents that aim only to achieve broad distribution. The interests and confrontations between these two approaches are very diverse (Van Dijck & Nieborg, 2009). The Web 2.0 (i.e., sharing social networks) has made a mistake by attempting to commercialise contents created by users (Jenkins, Ford & Green, 2015). 'While today's media environment is characterised by tailored media products, global media conglomerates, deregulation, flexible work arrangements, casualisation of the labour force, and increased consumer surveillance, these changes are extensions of earlier historical processes more than a radical break with the past' (Havens & Lotz, 2012: 199). What has changed are the participatory capabilities of users. It is true that 'we are at the beginning of a revolution that is fundamentally changing the way we live, work and relate to one another' (Schwab, 2017: 13). Yet power should involve achievement of adequate balance between large media conglomerates and the participation of people on the Internet.

In spite of technologies and lifestyles that encourage the individualisation of almost everything, new ways of belonging and creating community linked to the so-called transmedia culture are appearing. Moreover, all of this intersects with a new freedom of power to choose at each time and in each place because the decisive factor is access. The motto of this age could be 'anyone, anywhere, anytime' (Álvarez-Monzoncillo, 2011), giving rise to a new empowered consumer (Füller, Mühlbacher, Matzler & Jawecki, 2009).

This participative culture doesn't have any precedents. Young people embrace the Internet because they can find things that they love and it allows them to self-confirm themselves. They tend to get pulled in by a magical attraction and perhaps embrace the 'Californian Utopia' too quickly (Barbrook & Cameron, 2001), in contrast with critical capabilities when it comes to using mimetic reproduction models. In fact,

> The cyber-utopianism updates an idea very much present in the modern revolutionary movements: overcoming the traditional community guardianship and the appearance of a type of social interaction that is simultaneously both caring and respectful of individual freedom.... And promises us new digital lands, albeit perhaps too utopian. (Rendueles, 2013: 121)

Traditionally, in our societies, property is power. Yet change is also under way on that front. Sharing without owning homes, cars, computers, bicycles or energy implies the retirement of Adam Smith, as Rifkin (2011) proclaims. There are some shattered dreams hiding behind the participative culture. There is a lot of criticism regarding what is work and what is free, motivated by the sensation that free creators should work for the prestige and not for money (Fuchs, 2015). In the view of this author, who has a Marxist focus, the limits of the categories of industrial capitalism such as work time/leisure time, production/distribution/consumption and office and factory/home and privacy have begun to be more porous. The goal of capitalism is to intensify and spread the exploitation of workers via dualistic characteristics of Fordist capitalism (Fuchs, 2015).

The dynamics of individualisation of the Internet and the ability to participate brought the rhetoric of empowerment and the digital revolution with them, remixed with utopias involving social change and with other types of representation and/or democratic participation. Yet this empowerment capability has two sides: the empowered individual and the disempowered individual. The former feels that he/she has more power because technology allows him/her to become more informed and communicate and organise him/herself better, whilst the latter feels excluded and stripped of all power. It's the same old debate between the elites and the common people. There are two categories: the very active who are experiencing new ways of participating in civic life (underpinning the revival of the concept of citizenship) and those who have no influence and no organisational capabilities and are silent.

Effectively, there are contributions concerning social networks and the public sphere that may be a bit too optimistic regarding social change and that distance themselves from the approach of the concept developed by Jürgen Habermas (1992), who highlighted that the private ownership of the platforms that support and lead the transformation must be taken into account in any analysis. Of course there are other transformative forces to be discussed: social networks allow all citizens to change their relationship with the public sphere (Benkler, 2006); the emergence of a new sphere 2.0 allows citizens to participate and influence the public agenda (Papacharissi, 2009); the communication protocols that must be created are leading the construction of this new public sphere (Castells, 2011); and the platforms, like YouTube, as enablers of a new cultural public sphere, as they allow anyone connected to interact with different cultures, impacting beliefs and identities (Burges and Green, 2018). All this contributions demonstrate that empowerment involves more factors than merely having access to technologies that allow for more efficient communication and greater organisational power than others (Fuchs, 2015: 315).

The digital revolution has given more power to the people. The hierarchical organisation of flows of authority and power is something that belongs to the past. Political systems are also changing to adapt to new ways of organising ourselves and of disseminating opinions. However, the new power can control or even lead to battles in order to obtain programming (Rushkoff, 2010).

4.3 The business model era

Until the 1990s, the term *business model* was hardly ever used (Osterwalder, Pigneur & Tucci, 2005). However, this term quickly gained prominence among both practitioners and business scholars with the development of ICT and the emergence of Internet companies. As a result, it was necessary to analyse a new set of companies that were beginning to create markets and no longer resembled traditional industries (DaSilva & Trkman, 2014). In fact, most business models are currently affected in some way by technology (Kinder, 2002).

A general definition of the concept has not emerged, but it can be said that there is agreement among scholars that a business model must link the workings inside the firm to outside elements, including the customer side, and indicate how the value the company creates is captured or monetised (Baden-Fuller & Mangematin, 2013). In short, we can say that a business model 'defines how the enterprise creates and delivers value to customers, and then converts payments received to profit' (Teece, 2010: 173). It performs two important functions: value creation and value capture (Chesbrough, 2006).

Foss & Saebi (2015: 3) argue that business models are all about "How is it being done?" rather than "What is being done" and "How is revenue being captured?". The underlying argument is that business models uniquely address "how" issues, whereas the other issues are treated in the extant body of literature on marketing and competitive strategy. "It" refers to a particular business model. Business models then must design the organisational structures that a company needs in order to take advantage of a commercial opportunity (Foss & Saebi, 2015). However, this is not the same as a strategy. In fact, all companies aim to put some type of business model into place, but not all companies have a strategy (DaSilva & Trkman, 2014). Strategy is about building dynamic capabilities aimed at responding efficiently to future and existing contingencies. Its purpose is to successfully compete, whilst a business model aims to create value with the effective coordination of business resources (Osterwalder et al., 2005). To wrap up, strategy 'reflects what a company aims to become, while business models describe what a company really is at a given time' (DaSilva & Trkman, 2014: 383).

There isn't a unified typology of business models either (Foss & Saebi, 2015). Yet consideration of fundamental archetypes is important because of the features of market economies where there are consumer choices, transaction costs, heterogeneity amongst consumers and producers and competition (Teece, 2010). Broadly speaking, with regard to the media, we would mention participatory models, where the contribution of users is vital for value creation; distribution models, where contents must be created and revenues generated in several ways; and editorial models, where all content is offered in exchange for payment and free offerings are a rare exception (Lyubareva, Benghozi & Fidele, 2014).

Participatory models tend to mix, in particular, user-generated content with that from external producers. Value creation is dependent on users' contributions, so that the presence of free content is vital for the model (Anderson, 2010). Consumers have free access, so that it is necessary for audiences to be sold to advertisers. Thus, companies thereby become multi-sided platforms that match users with advertisers (Evans & Schmalensee, 2016). As they are closely linked to advertising, they are overly dependent on economic cycles. This is the case with groups such as YouTube and Facebook. Only those users contributing more successful content receive any economic compensation, which may lead to motivational problems over the long term. Costs are thereby reduced by substituting consumer input for editorial decisions. Content openness, combined with sharing and peering, have become the keys to the success of collective creativity (Lyubareva, et al., 2014). However, there are multiple offerings for commercialisation of contents (streaming, downloading, desktop, mobile, etc.).

As for distribution models, they tend to target a precise market segment and develop original content in-house. However, they also purchase content in the market that they cannot or do not want to create. They use several channels to generate revenues, such as public funding, donations, advertising, direct sales and/or subscriptions. These sources can be mixed in different proportions. In the case of usage of advertising for financing, they also become multisided platforms. They utilise multiple distribution channels to reach their consumers and must control final monetisation with them. Free-to-air TV, a free online newspaper and a newspaper that puts a paywall into place are some of the options within this general model.

Finally, editorial models tend to be based on content from external professional suppliers (although they may also have a small amount of in-house production). All content is offered in exchange for payment, and there are normally mechanisms in place to restrict usage (DRM or a similar approach). These companies must control the quality of their content and tend to structure price discriminatorily. Pay TV functions under this model.

Managers seeking to outperform their competitors should focus on choosing the right business model, implementing it as best as they can, improving the company's dynamics capabilities and engaging in business model change when an opportunity or threat arises (DaSilva & Trkman, 2014: 383).

The business model concept is gaining traction in different disciplines, but it is criticised for being vague and lacking theoretical foundations (Zott, Amit & Massa, 2011). Nevertheless, business models can be used to categorise the corporate world and explore the economic possibilities offered by different archetypes.

4.4 Mainstream media, niche markets and globalisation

The disruption of the Internet has underpinned an economic crisis that endangers the very survival of mass communication media, which triumphed in the 20th century in all countries. However, change has moved us towards the so-called matrix era, which is 'characterized by interactive exchanges, multiple sites of productivity, and diverse modes of interpretation and use' (Curtin, 2009: 13). While today's media environment is

> characterized by tailored media products, global multimedia conglomerates, deregulation, flexible work arrangements, casualization of the labour force, and increased consumer surveillance, these changes are extensions of earlier historical processes rather than a radical break with the past. Moreover, these changes are fundamentally tied to economic change, not to technological changes such as the growth of Internet access and digitisation, though these changes certainly helped facilitate the economic changes we are witnessing. (Havens & Lotz, 2012: 199)

The concept of television is transforming towards new forms and concepts such as OTT and TV-like services, between linear and nonlinear (Ala-Fossi, 2017).

Large conglomerates dominated the 20th century (Holt, 2011; Kunz, 2007). Their dominance has continued into the present even though the Internet is threatening the status quo (Álvarez-Monzoncillo & López-Villanueva, 2014). They imposed their domination around the world. Cultural imperialism appeared, mostly as a concept that has its foundations on the diffusion of foreign cultural values and beliefs, imposed by dominant leaders in the incumbent cultures (Tomlinson, 1991). Research from various different points of view has been conducted on this topic (Chen & Morley, 2006; Chomsky, 2010; Golding & Harris, 1996; McChesney, 2001; Mattelart, 1995; Morley, 2006; Schiller, 1971; Tomlinson, 1991; Wasko, 2013). This concept and related theories have surged on the back of globalisation and the rise of the Internet.

Yet these changes in the media follow certain trends and call for the reform of their business models and social roles (Freedman, Obar, Martens & McChesney, 2016). These trends have also been reinforced by neo-liberalism, and as a result, there has been a return of the state and/or the need for new regulations (Flew, Iosifidis & Steemers, 2016; Simpson, Puppis & Van den Bulck, 2016). Other future new consequences are also foreseen: 'great fragmentation, cheaper and cheaper production costs, the development of business models based on spatial rather than temporal distribution windows, and increasing surveillance of users to counter problems of fragmentation' (Havens & Lotz, 2012: 222).

However, the Internet also offers opportunities for distributing and accessing most audiovisual products, and the vast number of Net-enabled devices now available has transformed the audiovisual industry of the analogue era. The old status quo has been profoundly restructured. Those who own content want to distribute it directly to avoid paying an intermediary. Those who transport it on telecommunications networks want to obtain it to vertically integrate their activities and undermine Net neutrality.

> The big Internet players (Google, Amazon, Yahoo, iTunes, etc.) want more power and to defend their leading positions. It is a war between the multimedia convergence players. However, the emerging value chain looks unlikely to generate much revenue, and cannibalisation is still progressively questioning the analogue model with its barriers to entry and walled gardens. Meanwhile, audiences are fragmenting and new multi-device consumers are fickle and demand greater participation. (Álvarez-Monzoncillo, 2011: 195).

Whatever the case may be, it is yet to be seen who will control mainstream global entertainment versus regional blocs such as Europe, India, Latin America and China. This sector is controlled by North American companies that have adapted to new situations and continue to increase their exports. We are witnessing the dominance of

> hip capitalism, a new advanced global cultural capitalism, at the same time both highly concentrated and very decentralized ... that is constantly transforming, permanently adapting, as the creative industries are no longer factories like the studios of Hollywood's Golden Age, but rather production networks comprised of hundreds of small emerging companies ... whose results are increasingly asymmetrical North-South exchanges and increasingly unequal South-South exchanges among emerging and disadvantaged countries, yet still dominated by an increasingly powerful country. (Martel, 2010: 419)

This author foresees the ongoing dominance of North American entertainment and culture in the present century in spite of substantial fragmentation and the appearance of new market niches, reinforced by the Internet. We should also bear in mind that we have observed the slight emergence of the Asian continent (Curtin & Shah, 2010). Transmedia cultural is being born, and 'glocal' strategies (global mainstream with local culture) are here to stay. Moreover, it is still too soon to predict the death of the mass media of the 20th century. The long-tail theory does not appear to have held true in a market that is still dominated by hits (Elberse, 2013).

In short, the trends seem evident. Firstly, the dominance of the large North American conglomerates of the entertainment industry should be similar to that of the 20th century in the upcoming decades. The companies making up the rest of the industry will hardly exit their regions or zones of influence (e.g., the European Union) or linguistic areas (e.g. Spanish).

Secondly, it doesn't look like we are set to face the end of the 'couch potato' model. Long live television! Nevertheless, the former model coexists with new à la carte consumption because users want to choose. The slogan 'anyone, anytime, anywhere' on different platforms and devices is an indisputable fact. As a result, audiences are fragmenting. Thirdly, user-generated content will be very important, and there will be conflicts between companies and users when it comes to monetising these contents. Fourthly, personal information and data analysis will determine new uses and marketing strategies of the media and entertainment industry. Lastly, the market will continue to be dominated by blockbusters, whilst niche products should register slower growth than expected with the development of the Internet and globalisation.

4.5 Privacy and security: Personal information as a commodity

George Orwell's prophecy in *1984* doesn't seem too far out. In all types of societies and systems, different forms of control have existed. Power doesn't exist without control. Deleuze introduced the concept of a 'control society' in 1990s. Beck's concept of a risk society doesn't seem too far out either (Beck, 1998). Currently, people share personal information and thereby create a digital identity. Some people manage this process and others don't, but, in general, the basis is the control of social relationships. With the boom of social relationships carried out via electronic devices, personal information has become a business that is on the rise.

Whilst companies and states are collecting information regarding people in order to design strategies and guarantee security within a global scenario, data protection and privacy are being demanded as inalienable rights. Cyber control is intimately related to the information society and the transparency of relationships.

Moreover, surveillance is carried out automatically and anonymously. 'Likes' are converted into business, whilst automatic calculations track online activities in order to identify and gain knowledge about profiles that are later used to make marketing decisions. As a result, 'digital industrialism turns users and their personal data into a new commodity' (Rushkoff, 2016: 44).

With the proliferation of connected devices, social networks and the so-called Internet of things, private lives tend to be diluted. This trend is reinforced by implants of chips or devices that allow for geolocalisation. As a result,

> Videosurveillance linked to a database allows for computer analysis of behaviour. Multiple digital technology applications make available a massive amount of information regarding people, whilst there are an increasing number of procedures for usage of this data. Virtual oceans of data are the object of all kinds of treatment in unknown and inaccessible places. The miniaturisation and the dematerialisation of IT

hardware increases the invisibility of surveillance. The current state of IT in the cloud leads to thoughts of dematerialised information technologies, where applications and data are consulted remotely without the need for any local infrastructure. (Mattelart & Vitalis, 2014: 197)

Thus, the 'panopticon' model, a place from which everything can be seen—conceived by Bentham and by Foucault's research regarding the balance of power between the watched and the watchman—is making way for the post-panopticon model proposed by Bauman when he states that 'liquid surveillance' is less a complete way of specifying surveillance and more an orientation, a way of situating surveillance developments in the fluid and unsettling modernity of today (Bauman & Lyon, 2013: 10). This model involves not just surveillance, since 'forms of control that group together very different perspectives have appeared. Not only do they have an obvious connection with the idea of imprisonment, but rather they also frequently share characteristics of flexibility, fun and entertainment, and consumption' (Bauman & Lyon, 2013: 13). This concept is quite aberrant if we think that whilst a person is playing and having fun, information can be obtained that will allow his/her preferences and potential behaviour to be predicted.

Technology allows for 'inverse surveillance'. The watcher watches over the watchman. Steve Mann invented the word *sousveillance*, or surveillance from below (mentioned by Mattelart & Vitalis, 2014: 205). These authors develop Siva Vaidhyanathan's theory of the 'nonopticon' model, which is characterised by not knowing who is watched and who is watching or with what level of indiscretion. It appears to be the perfect situation for both parties.

It is this hiding of surveillance that is making young people extremely and unknowingly vulnerable as personal histories are being created and may turn out to be very interesting for a lot of intermediaries that will thereby be able to monetise their digital fingerprints. These entities know what young people are looking for on the Internet, what they share, what they exchange and what they like. All of this is underpinned by the so-called selfiemania. Thus, selfies can be considered a metaphor of the 'humanised network', with people aiming to pursue fame within a social network. They are simultaneously inputs demanding a certain level of interactivity with other interlocutors who can make future comments. Additionally, they are susceptible to becoming 'viral images', as an ordinary event of a personal and, to some extent, quite private nature may end up becoming a viral prototype: exteriorising certain intimacies and fragmenting or transcending filters in order to expand towards other circuits and networks for which they weren't initially intended (Gómez-Alonso, 2016: 20).

In short, they are watched narcissists. Moreover, Instagram can be considered an application used to create 'personal brands'. We are our own agents, now capable of efficiently 'advertising oneself', in a process defined by Michael Moritz, chairman of Sequoia, as 'the personal revolution'

(Keen, 2016: 150). We are facing a narcissistic epidemic that Richard Brooks calls 'expressive individualism'. Instagram is not the only social medium that has crossed the line of narcissism; 'Twitter, Tumblr, Facebook and the other social networks, apps and platforms that feed our vain "selfiecentric" illusions in the middle of an apparently infinite hall of mirrors have also done so' (Keen, 2016: 153).

Narcissism and surveillance seem to understand each other. The idea of sharing your state of mind via photos and/or what you are thinking at any moment in time feeds this need to socialise. And there are companies that are willing to pay and charge for this information. They know you better and know what you want. It's the dream of any salesperson and without any intermediaries. It is a new scenario where either you program them or they program you (Rushkoff, 2010).

4.6 Talent and labour

The impact of the Internet and the digital economy (Tapscott, 1997) on the job market has been studied from several perspectives (Mesenbourg, 2001). Being a disruptive technology, the displacement of employees began at the very beginning. Traditional industries based on information like media, music, film, television and games were the first to suffer the consequences. These effects went unnoticed due to the sharp economic growth that was generated at the same time in the global economy. Maybe knowledge workers were not, as predicted (Drucker, 1999:79), 'the most valuable asset of a 21st-century institution (whether business or non-business)'.

Nevertheless, the problem of productivity and technology (Brynjolfsson & McAfee, 2011) can be contrasted with the vision of how 'the provision of free labour is a fundamental moment in the creation of value in the digital economies' (Terranova, 2000: 36). Some others considered that 'the problem of quantifying the value of knowledge makes it difficult to solve the dilemma of whether the Internet has created or destroyed the labour market' (Álvarez-Monzoncillo, Suárez-Bilbao & de Haro, 2016: 264). Whilst it is clear that 'machines are substituting for more types of human labour than ever before' (Brynjolfsson, McAfee & Spence, 2014: 44) and robots are taking over, we cannot take for granted what a Toyota manager explained: 'Robots don't make suggestions' (Pine, Victor & Boynton, 1993: 108).

Marc Andreessen, Netscape founder and investor in technology, stated that 'software is eating the world', mainly turning the labour market into a hollow trench where 'the spread of computers and the Internet will put jobs in two categories: People who tell computers what to do, and people who are told by computers what to do' (Andreessen, 2011). More and more new companies are created based on software searching to disrupt the traditional models. Many of them are helped by incubators like the Y Combinator (Stross, 2011), where in three months no more than three founders, mostly software developers or hackers, create for the Demo Day a prototype and raise money to grow start-ups that are reshaping the business world, like Dropbox or Airbnb.

The value chain has been turned upside down. 'Companies no longer design, make and then sell products; instead, companies will sell capabilities, get orders, and then fulfil these requests. Consequently, their success will depend very much on the ability to manage knowledge' (Tseng & Piller, 2011: 10). This requires a new type of worker: 'Gone are the days of the traditional 9-to-5. We're entering a new era of work—project-based, independent, exciting, potentially risky, and rich with opportunities' (Horowitz; & Rosati, 2014). Not only are new skills needed, but also life in general will be completely adapted to the new competitive and social environment.

The impact of robots on professions will affect not only the ones previously mentioned but also healthcare services (diagnostics or triage is better with machines), education (MOOCs, or massive open online courses), religion (apps with access to whole sacred texts, virtual reality to pray and pilgrimage), law (ODR, or online dispute resolution) and more (Susskind & Susskind, 2015).

More and more the labour market is going global. Unemployment is rising, and 'work for life' is no longer a goal for young people. Talent management and development is key for competition.

4.7 At the doors of the Fourth Industrial Revolution

The global industrial map has become increasingly complicated over the last 40 years, with major reconfigurations, both qualitative and quantitative, of former ways of manufacturing, distribution and consumption (Dicken, 2011: 14). Direct foreign investment, outsourcing, delocalisation and cross-border trade have risen substantially throughout this period.

All of the above has been supported by an increase in the industrial capacities of developing countries and the improvement of their institutions, lower transportation costs, the end of the Cold War and the rise of market systems, lower customs tariffs and new ITCs that allow for integration of remote activities (Mankiw & Swagel, 2006: 1053). This has created the dichotomy of a world that is shrinking by connecting regions that were historically distant but that, at the same time, is getting bigger due to expansion of trade horizons (Osterhammel & Petersson, 2003: 3).

Thanks to these driving forces, companies have been able, gradually, to slice up and relocate their supply and distribution chains. This has given rise to industrial activities that are fragmented and disperse from a geographic point of view but integrated from an economic point of view. The division of labour has thereby reached a new dimension as work has been disaggregated into a greater number of activities in different places with complex connections. This unbundling has transformed the global economy in different ways under a process that is still ongoing (Baldwin, 2012: 7).

There are three constraints that hold back the globalisation of markets: the costs of moving goods, ideas and people (Baldwin, 2016: 8). In the pre-globalisation world, all of them were bundled together to such an extent that the world economy was 'a patchwork of village-level economies' (Baldwin, 2016: 4). The First Industrial Revolution allowed the costs of transporting physical goods to decline, thanks to the steamship and the railway. This process, which began around 1820, made the separation of production and consumption—globalisation's first unbundling—possible. The Second Industrial Revolution, with technologies such as the telegraph, electricity and assembly lines, accelerated the dynamics of this trend at the end of the 20th century. From 1990, ITCs radically drove down the costs of moving ideas. This second unbundling process underpinned the international separation of factories via a global value-chain revolution.

Since the 1970s, we have witnessed the deindustrialisation of the west and the industrialisation of the so-called emerging countries, with China at the top of the list. This has reduced the divergence of revenues between hemispheres that had been taking place since the First Industrial Revolution, allowing for new patterns of consumption in emerging countries thanks to their rapid economic growth. Countries are now able to industrialise by joining these new supply chains, and they don't need to build up a broad industrial base in order to be competitive anymore. Finally, trade is no longer limited to final goods crossing borders; rather, the proportion of intermediate goods or products in the process of being manufacturing is rising. In fact, currently, nearly 80% of global trade involves networks coordinated by large companies (UNCTAD, 2013: 135). Moreover, international investments in factories, technologies, training, marketing and intellectual property are intertwined. All of this process is underpinned by services that coordinate this dispersion such as telecommunications, the Internet, express couriers and container ships.

Major development of ICTs has led some authors (for example, Freeman & Louçã, 2002: 301) to talk about a Third Industrial Revolution. These technologies are increasingly powerful, multifunctional and online, although in recent times it doesn't look like they have managed to generate substantial productivity improvements. The great influence that they had on the most tedious administrative tasks ended in the 1980s, and in the new century, innovations have focused on entertainment and communication products, which are increasingly smaller and more intelligent but don't spur changes in labour productivity or lifestyles, unlike the effects of electricity and the automobile in their day (Gordon, 2016).

Digital and physical worlds are forecast to become inseparable in the Fourth Industrial Revolution (Brynjolfsson & McAfee, 2014; Schwab, 2017). Three-dimensional printing, DNA films, gene editing, robotics, artificial intelligence, the Internet of things and many more emerging trends could form an unparalleled melding of the physical, biological and digital worlds (Schwab, 2017). According to Brynjolfsson and McAfee (2014), whole categories of work will be transformed by the power of computing and the impact of robots.

The present clearly offers a resurgence of automation anxiety, but the basic fact that is often forgotten is that technology eliminates jobs, not work (Autor, 2015). In fact, we tend 'to overstate the extent of machine substitution for human labour and ignore the strong complementarities between automation and labour that increase productivity, raise earnings, and augment demand for labour' (Autor, 2015: 5). Moreover, in the case that human labour was rendered superfluous by automation, the great economic problem that we would face would involve the distribution of everything produced rather than scarcity.

A technology (or the artefact to which it gives rise) has no power in and of itself; it does nothing. Only in combination with people and social structures can technologies meet their goals. In other words, we are not dealing with technology per se but rather with technology within a certain context. As always, the social and the technological must be combined.

4.8 The complex boundaries of sharing

The word that describes our participation in social networks—and that has quite a nice ring to it—is *sharing*. This word seems to imply notions of equality, giving, collaborative consumption and sustainability. However, the concept of sharing is an undertheorised one and includes several differing logics between which it is necessary to make a distinction (John, 2012).

Generally speaking, we can make a distinction between two types of acts within the concept of sharing: distribution and communication. The act of sharing is one of concrete distribution when we physically divide something up between several recipients. In this case, it would be a zero-sum game, like when a child shares a candy bar and ends up with less, as governed by cultural norms (John, 2013). There can also be concrete sharing situations where the product doesn't run out, such as when a dorm room is shared or photos, movies and/or links are circulated. Finally, there is abstract distribution—in a way that is not a zero-sum game—when we have something in common with others (e.g., a belief can be shared between people). Sharing can also be an act of communication, such as when we talk about sharing our emotions or when we update our Facebook status (John, 2013).

Sharing material things tends to require some form of sacrifice on the part of the person sharing, whilst with immaterial things nothing is reduced. However, there may be infringement of intellectual property rights of a third party, thereby reducing his/her potential revenues. In the predigital age, sharing was always mutual, social and based on the principle of reciprocity (Wittel, 2011).

The sharing of digital things does not involve any sacrifice, as there does not tend to be any rivalry or exclusion. Under this scenario, there is an extension of ownership rather than a transfer of ownership. Moreover, transaction and transportation costs are minimal (Henten & Windekilde, 2016), which may allow for a global reach. Often, on the Internet, distribution and communication go hand-in-hand: if we share photos of a trip, we are communicating our lifestyle. Nevertheless, while digital content may be dematerialised, limitless and with zero marginal costs of use (Rifkin, 2014), it is still closely coupled with tangible products, and the resources that make it possible (such as electricity, servers and bandwidth) are limited (Kennedy, 2015).

Within this 'maze of terms' (Belk, 2014), it is difficult to discern where sharing ends and commerce begins. Money transforms the sharing transaction into a commodity exchange. The sharing economy concept is said to build on the concept of collaborative consumption (Hamari, Sjöklint & Ukkonen, 2013). Nevertheless, the concept of collaborative consumption, as it was first put forward by Felson and Spaeth (1978), had a different meaning—namely, 'events in which one or more persons consume economic goods or services in the process of engaging in joint activities'. The examples were 'drinking beer with friends, eating meals with relatives, driving to visit someone or using a washing machine for family laundry are acts of collaborative comsumption' (Felson & Spaeth, 1978: 614). Actually, there is great dissonance between the sharing that many Internet companies claim to do and reality: they are 'attention merchants' that sell our data (Wu, 2016), whilst they exploit the free work carried out by users of these platforms.

The sharing economy is an 'economy in which the rental contract supersedes outright ownership transfer under a private property rights system' (Tsui, 2016: 80). There are sharing economies of production—such as Java, Linux or Wikipedia (Benkler, 2006; Lessig, 2006; Tapscott & Williams, 2006)—and sharing economies of consumption—such as Majorna, a small-scale neighbourhood-based car-sharing entity in Göterbor, Sweden, with no employees (Belk, 2014). Paradoxically, in its purest state, the sharing economy would be more of an 'anti-economy' than any economy at all. This reflects the fact that there are practices that cannot ultimately be monetized, which leads to their disappearance or to the conversion of this sharing into an exchange (Sützl, 2014).

In order to have a clearer view of the phenomenon of sharing and separate it from digital markets that pretend to be part of this trend, we should bear in mind that the ownership of consumer goods can take two basic forms: they can be owned individually or conjointly. It is also necessary to differentiate the motivations for owning specific consumer goods. These can be pecuniary and non-pecuniary. This is the approach of Maurie J. Cohen (2017), and with the juxtaposition of these two ownership features (type and motivation), we can distinguish four archetypes (refer to Figure 4.1).

Quadrant 1 encompasses those goods that that are not owned for pecuniary reasons and are private property, such as an automobile that is not used to generate any kind of revenue. In this case, sharing of distribution is possible if so desired by the owner, since all goods are potentially shareable (Wittel, 2011). In Quadrant 2, consumer goods are deployed for non-pecuniary purposes. This would be the case with the Majorna car-sharing club; the distribution of photos, links and videos by users of a platform, even though they are not the owners of the rights; fraternal organisations; and public

Ownership motivation	
Pecuniary	
4 Brokered micro-entrepreneurship (micro-entrepreneurs working for Uber...)	3 Serialised rental (Avis, Zipcar, Airbnb...)
Individual	*Conjoint*
1 Private ownership/usership (the owner decides)	2 Communitarian provisioning (Majorna club, distribution of photos, links...)
Non-pecuniary	

Figure 4.1 Provisioning archetypes

transportation. These are cases of pure sharing. As for Quadrant 3, it represents serialised rental such as Avis, Zipcar or Airbnb. In this category, we also find large Internet companies such as Facebook and Google, which connect users from Quadrant 2 with advertisers. Finally, Quadrant 4 contains micro-entrepreneurs that use their automobiles to transport others. Yet, in this case, there are pecuniary reasons, and there are more and more digital platforms for bringing together users and service providers. In other words, the upper quadrants can end up becoming classic commercial exchanges.

The Internet has opened up many new possibilities for sharing and commodity exchange, to the point that there has even been a 'war on sharing' (Aigran, 2012) waged by many industries that have seen their business models collapse. Nevertheless, minimal true sharing seems to be taking place; rather, the majority correspond to new business models. Sharing is a major force that has been strengthened by the Internet but does not create a new social organisation system, does not reduce the expenditure of resources, does not foster economic stability and does not increase diversity.

4.9 Mobility

Human computer interaction suffered an inflexion point with the spread of mobile computing. What began as an increase in dedicated hardware and networks turned into a business of devices and software (apps) via online digital shops. The mobile world is changing the way we socialise, create and manufacture goods. In the last decade of the previous century, telephone service providers focused on voice services and later experienced the advent of smartphones and the disruption in the category created by iPhones (Vogelstein, 2008). Precursors of the Internet of things and creators of the need for Big Data systems, we could not have created an 'anywhere, anyhow, anytime' world without them.

Understanding that the 'Internet of Things refers to the networked inter-connection of everyday objects, which are often equipped with ubiquitous intelligence' (Xia, Yang, Wang & Vinel, 2012: 1101), we can see how the increase in connected nodes has led to an increase in the volume, speed and variety of data needed to understand the world today. Intelligence, which was traditionally held and controlled in the network, is now moving towards the nodes. This shift in power has an influence on all aspects of current life, from social to economic to demographic. Even the legal system is struggling to keep pace with the current situation (Weber, 2010). The possibility of machine-to-machine environments that can create data monopolies (Holler et al., 2014) may change the way companies and countries compete in the future. The Internet of things is also a driver of the creation of wearables (Wei, 2014), which have moved from devices with one main function to those with multifunctional capabilities that generate a higher amount of personal data for an industry. Regardless of divergent forecasts carried out by inter-ested parties, this is 'typical when industries are in their relative infancies and hypergrowth mode' (Wei, 2014: 53).

Another relevant impact of mobility is mass customisation, which can be defined as 'developing, producing, marketing, and delivering affordable goods and services with enough variety and customisation that nearly every-one finds exactly what they want' (Pine, 1993: 44). Tseng and Jiao (2001: 685) updated this definition as 'producing goods and services to meet individual customer's needs with near mass production efficiency'. The idea of produc-ing where we need the goods is not new. Nevertheless, ICTs brought this possibility closer than ever. Anderson (2012) explains how a new generation of 'DIYers' (do-it-yourselfers) may find all the information they need in com-munities of co-creation while manufacturing locally with 3D printers.

This manufacturing system is also creating great value for traditional companies. GE has created an engine with 3D printing with several advan-tages. The first one is that the previous traditional version needed several components, which were manufactured in different places and then com-bined (Ford, 2016: 173). This system reduces the cost, time and probability of errors. In addition, the weight of the final engine is reduced, which has great implications for lowering costs of fuel usage and, consequently, lead-ing to cleaner operations. In the end, 'Mass customisation and customer integration create a customer centric enterprise system that transcends the traditional manufacturing enterprise' (Tseng & Piller, 2011: 10). Anywhere, anyhow, anytime at its best.

The credit for coining the term *Big Data* is given to John R. Mashey due to a presentation made in April 1998 titled 'Big Data ... and the Next Wave of InfraStress'. How should we deal with this huge amount of data that is now diverse, is available in greater volumes than ever before and is created in real time? Social media moguls like Facebook leveraged on the problem that 'Businesses are creating more data than they know what to do with' (McAfee, Brynjolfsson, Davenport, Patil & Barton 2012: 61). The time we dedicate today to traditional media is split among social media companies, mobile devices and companies like Google that have created new services in order to solve this huge content problem.

The next step in the process is the autonomous car, a source of personal data but also a change in 'the machine that changed the world' (Womack, Jones & Roos, 1991). It should lead to fewer accidents, better usage of cars and a cleaner environment. The impact will affect more than just the automotive industry; rather, certainly, there will be again unexpected changes in economics and labour to come (Ford, 2016).

4.10 Social revolutions

In the Web's beginnings, 'content creators were few in Web 1.0 with the vast majority of users simply acting as consumers of content' (Cormode & Krishnamurthy, 2008). The advent of Web 2.0 increased interactivity, interoperability and usability for end users, who began to participate and produce user-generated content. This concept was coined in the 20th century (DiNucci, 1999) but popularised by Tim O'Reilly and Dale Dougherty after a conference in 2004. The idea of the Internet as a more user-friendly platform brought a brand-new world of content and interaction.

Search was mandatory to navigate through growing volumes of content, bloggers proliferated and the syndication of content permanently changed the media world. Search engine optimisation became the new religion of publishers and the cost per click its exchange coin. Wikipedia's success demonstrated how collaborative business models were set to become the new normal (de Haro & Cereijo, 2016) not only for technical projects (like Linux) but also for content ones. The impact of mass collaboration, thought to be disruptive for coding and content businesses, is now affecting several others, such as mining: e.g., GoldCorp in Canada was facing a severe drop in production that it solved via collaborative methods (Tapscott & Williams, 2008).

The Amazon Mechanical Turk was an online platform on which employers could post 'human intelligence tasks' to be selected by 'prosumers' (Toffler, 1980) who performed only their preferred knowledge tasks at their homes, submitted results and then got paid only for the final output. Today, mobile apps in Sweden reshape sales-force management with a similar production process.

The social web (Rheingold, 1996), understood initially as the set of relations that link people online, turned into the next big thing of current generations. This traced back to basic bulletin board systems, which operated like discussion forums. The creator of the Internet, Sir Tim Berners-Lee, stated, 'The Web is more a social creation than a technical one' (Berners-Lee, 1999: 123). Communities led the transformation. Companies like Classmates.com (1995) and SixDegress.com (1997) were pioneers, whilst MySpace (2003) and Facebook (2004) grew the flame. Dunbar's number (Dunbar, 2010) was challenged and validated by Twitter and other social networks. The networked society led to new social movements, as defined by Castells (2011, 2015).

4.11 Generation gap

We understand that the meaning of *generation* is a group of people of similar ages who share common experiences (Rudolph & Zacher, 2017). However, the problem with generational typologies is that their grain of truth is underpinned by the undisputed fact that there are certain similarities among members of heterogeneous population groups but that they are stereotyped by their year of birth (Fineman, 2014). Once that typology and the name given to a generation take hold (particularly in the media, consulting and marketing), it is very difficult to shake it off, and it tends to be considered to be an undisputed fact.

The great dilemma that is involved when speaking about a generation consists of separating the effects of three related but very different factors: age, historical period and statistical cohort. The difficulty lies in how to determine the variance of each of these variables independently from the others, as a generation tends to be seen as an intersection of age and period that gives rise to a group of individuals with shared experiences (cohort). Based on this view, results are inexorably intercorrelated (Costanza & Finkelstein, 2015).

As pointed out by Paul Sackett (2002), it is extremely complicated to use age and historical periods in order to compare groups of people. It is necessary to specify the events and experiences that underpin the hypotheses regarding differences between cohorts and to systematically test these hypotheses. If this is not done, we can end up inventing stereotypes.

First studied by Mannheim (1952), generational gaps are based on the idea of a group of people sharing the same social and historical locus. Under those circumstances, environmental conditions may lead to a common interest in differentiation from previous generations.

As a result, stereotypes act like cognitive shortcuts that save us time, as they allow us to make quick judgements in a complex world. Thus, we tend to store and record information regarding a group that is consistent with the stereotype and to discard those examples that are not in accordance. We look for shortcuts such as 'if they are "Baby Boomers" they will do X and if they are "Millennials" they will do Y', but generalisations regarding groups tend to be discredited over the long term (Costanza & Finkelstein, 2015: 313).

Treating the members of a generation as if they were intrinsically and uniformly similar hides the fact that each person has his/her own desires, talents, preferences and attitudes. In fact, acting this way conflicts with what we know about individual differences (Sackett, 2002).

There is little empirical evidence supporting the existence of differences based on generational circumstances, and there aren't any theories backing it up (Costanza & Finkelstein, 2015). In fact, there are many viable alternative explications of observed differences.

For example, millennials, who are acknowledged as 'digital natives' (Ransdell, Kent, Gaillard-Kenney & Long, 2011), are supposed to have some advantages: superior technological capabilities, greater abilities for dealing with scenarios of rapid and ongoing changes, a higher level of independence and better innovation capabilities than previous generations (Tapscott, 1997), as well as a remarkable combination of gearing to success and self-absorption (Zemke, Raines & Filipczak, 2000). Nevertheless, this may merely reflect the ways of behaving among young people during any particular historical period: they tend to be lazier, more energetic, more exploratory, more selfish and more dramatic than are their elders (Steel & Kammeyer-Mueller, 2015). Of course, these are only group trends and do not reflect the huge variations within each group.

In order to lead the new industrial revolution, there is a new generation that is better prepared than any other, which leads us to contemplate the problems of the generational gap and the digital divide (Gravett & Throckmorton, 2007; Howe & Strauss, 2009; Tapscott, 2009; Van Deursen & Van Dijk, 2014). Many will miss the train of digitalisation of the new society that we are building (Friemel, 2016). However, this trend reflects the existence of gaps that are present within a generation for economic, cultural and educational reasons. Moreover, there are also other perspectives regarding the so-called Generation Debt (Kamenetz, 2006).

References

Aigrain, P. (2012). *Sharing: Culture and the economy in the Internet age*. Amsterdam: Amsterdam University Press.

Ala-Fossi, M. (2017). Defining 'TV-like' content in a multimedia era'. DOI: 10.1093/acrefore/9780190228613.013.191

Álvarez-Monzoncillo, J. M. (2011). *Watching the Internet: The future of TV*. Lisbon: Media XXI.

Álvarez-Monzoncillo, J. M., & López-Villanueva, J. (2014). Barbarians at the gates of the cultural industries: Three possible scenarios. In P. Faustino, E. Noam, C. Scholz & J. Lavine (Eds.), *Media Industry Dynamics: Management, Concentration, Policies, Convergence and Competition* (pp. 497-514). Lisbon: Media XXI.

Álvarez-Monzoncillo, J. M., Suárez-Bilbao, F., & de Haro, G. (2016). Challenges and considerations of the new labor market in the media industry. *El Profesional de la Información, 25*(2), 262-271.

Anderson, C. (2010). *Free: The future of a radical price*. New York: Hyperion Books.

Anderson, C. (2012). *Makers: The new industrial revolution.* London: Random House.

Andreessen, M. (2011, 20 August). Why software is eating the world. *The Wall Street Journal, https://www.wsj.com/articles/SB1000142405311190348090457651225091 29460.*

Autor, D. H. (2015). Why are there still so many jobs? The history and future of workplace automation. *Journal of Economic Perspectives, 29*(3), 3-30.

Baden-Fuller, C., & Mangematin, V. (2013). Business models: A challenging agenda. *Strategic Organization, 11*(4), 418-427.

Baldwin, R. (2012). *Global supply chains: Why they emerged, why they matter, and where they are going*: Geneva: Center for Trade and Economic Integration, The Graduate Institute.

Baldwin, R. (2016). *The great convergence: Information technology and the new globalization.* Cambridge: Belknap Press.

Barbrook, R., & Cameron, A. (2001). Californian ideology. In P. Ludlow (Ed.), *Crypto anarchy, cyberstates, and pirate utopias* (pp. 362-387). Cambridge, MA: MIT Press.

Bauman, Z., & Lyon, D. (2013). *Vigilancia líquida.* Barcelona: Paidós.

Beck, U. (1998). *World risk society.* Cambridge: Polity Press.

Belk, R. (2013). You are what you can access: Sharing and collaborative consumption online. *Journal of Business Research, 67,* 1595-1600.

Belk, R. (2014). Sharing versus pseudo-sharing in Web 2.0. *Anthropologist, 4*(2), 7-23.

Benkler, Y. (2006). *The wealth of networks: How social production transforms markets and freedoms.* New Haven, CT: Yale University Press.

Berners-Lee, T. (1999). *Weaving the Web: The Original Design and Ultimate Destiny of the World Wide Web by its Inventor.* New York: HarperCollins Publishers Inc.

Brynjolfsson, E., & McAfee, A. (2012, 11 December). Jobs, productivity and the great decoupling. *New York Times. http://www.nytimes.com/2012/12/12/opinion/global/jobs-productivity-and-the-great-decoupling.html.*

Brynjolfsson, E., & McAfee, A. (2014). *The second machine age.* New York: Norton.

Brynjolfsson, E., McAfee, A., & Spence, M. (2014). Labor, capital, and ideas in the power law economy. *Foreign Affairs, 93,* 44.

Burgess, J., & Green, J. (2018). *YouTube: Online video and participatory culture (2nd Ed.).* Cambridge, UK: Polity Press.

Castells, M. (2007). Communication, power and counter-power in the network society. *International Journal of Communication, 1*(1), 238-266.

Castells, M. (2011). *The rise of the network society: The information age* (vol. 1). North Chelmsford, MA: Wiley.

Castells, M. (2015). *Networks of outrage and hope: Social movements in the Internet age.* North Chelmsford, MA: Wiley.

Chen, K. H., & Morley, D. (Eds.). (2006). *Stuart Hall: Critical dialogues in cultural studies.* London: Routledge.

Chesbrough, H. (2006). Boston: Harvard Business School Press.

Chomsky, N. (2010). *Hopes and prospects.* New York: Haymarket Books.

Cohen, M. J. (2017). *The future of consumer society.* New York: Oxford University Press.

Cormode, G., & Krishnamurthy, B. (2008). Key differences between Web 1.0 and Web 2.0. *First Monday, 13*(6).

Costanza, D. P., & Finkelstein, L. M. (2015). Generationally based differences in the workplace: Is there a there there? *Industrial and Organizational Psychology, 8*(3), 308-323.

Curtin, M. (2003). Media capital towards the study of spatial flows. *International Journal of Cultural Studies, 6*(2), 202-228.

Curtin, M. (2009). 'Matrix media'. In G. Turner & J. Tay (Eds.), *Television studies after TV: Understanding television in the post-broadcast era*. London: Routledge.

Curtin, M., & Shah, H. (Eds.). (2010). *Reorienting global communication: Indian and Chinese media beyond borders*. Chicago: University of Illinois Press.

DaSilva, C. M., & Trkman, P. (2014). Business model: What it is and what it is not. *Long Range Planning, 47*(6), 379-389.

de Haro, G., & Cereijo, M. (2016). Riesgos y limitaciones económicas del consumo colaborativo. *Harvard Deusto Business Review,* (260), 56-67.

Dicken, P. (2011). *Global shift*. London: SAGE.

DiNucci, D. (1999). Fragmented future. *Print, 53*(4), 32.

Dunbar, R. (2010). *How many friends does one person need? Dunbar's number and other evolutionary quirks*. London: Faber & Faber.

Drucker, P. (1999). Knowledge-Worker Productivity: the Biggest Challenge. *California Management Review, 41*(2), 78-94.

Elberse, A. (2013). *Blockbusters: Why big hits–and big risks—are the future of the entertainment business*. London: Faber & Faber.

Evans, D. S., & Schmalensee, R. (2016). *Matchmakers: The new economics of multisided platforms*. Boston: Harvard Business Review Press.

Felson, M., & Spaeth, J. (1978). Communitive structure and collaborative consumption. *American Behavioral Scientist, 21*(4), 614-624.

Fineman, S. (2014). Age matters. *Organization Studies, 35*, 1719–1723.

Flew, T., Iosifidis, P., & Steemers, J. (2016). *Global media and national policies: The return of the state*. London: Palgrave.

Ford, M. (2016). *El auge de los robots: La tecnología y la amenaza de un mundo sin empleo*. Barcelona: Espasa.

Foss, N. J., & Saebi, T. (2015). *Business model innovation*. Oxford: Oxford University Press.

Freedman, D., Obar, J., Martens, C., & McChesney, R. W. (Eds.). (2016). *Strategies for media reform: International perspectives*. Oxford: Oxford University Press.

Freelancers Union & Elance-oDesk. (n.d.). *Freelancing in America: A national survey of the new workforce.*

Freeman, C., & Louça, F. (2002). *As time goes by: From the industrial revolutions to the information revolution*. Oxford: Oxford University Press.

Friemel, T. N. (2016). The digital divide has grown old: Determinants of a digital divide among seniors. *New Media and Society, 18*(2), 313-331.

Fuchs, C. (2015). *Culture and economy in the age of social media*. London: Routledge.

Füller, J., Mühlbacher, H., Matzler, K., & Jawecki, G. (2009). Consumer empowerment through Internet-based co-creation. *Journal of Management Information Systems, 26*(3), 71-102.

Golding, P., & Harris, P. (Eds.). (1996). *Beyond cultural imperialism: Globalization, communication and the new international order*. London: SAGE.

Gómez-Alonso, R. (2016). Selfie forever: La edad de oro de la egolatría. En J. C. Alfeo & L. Deltell, *La mirada mecánica: 17 ensayos sobre la mirada fotográfica*. Madrid: Fragua.

Gordon, R. J. (2016). *The rise and fall of American growth: The U.S. standard of living since the Civil War*. Princeton, NJ: Princeton University Press.

Gravett, L., & Throckmorton, R. (2007). *Bridging the generation gap: How to get radio babies, boomers, Gen Xers, and Gen Yers to work together and achieve more*. Franklin Lakes, NJ: Career Press.

Habermas, Jürgen. 1992. *Further Reflections on the Public Sphere and Concluding Remarks In Habermas and the Public Sphere*, Calhoun, C (Ed.) (pp. 421-479). Cambridge, MA: MIT Press.

Hamari, J., Sjöklint, M., & Ukkonen, A. (2013). The sharing economy: Why people participate in collaborative consumption. *Working Paper*.

Havens, T., & Lotz, A. (2012). *Understanding media industries*. Oxford: Oxford University Press.

Henten, A. H., & Windekilde, I. M. (2016). Transaction costs and the sharing economy. *Info, 18*(1), 1-15.

Holler, J., Tsiatsis, V., Mulligan, C., Avesand, S., Karnouskos, S., & Boyle, D. (2014). *From machine-to-machine to the Internet of things: Introduction to a new age of intelligence*. Oxford: Academic Press.

Holt, J. (2011). *Empires of entertainment: Media industries and the politics of deregulation, 1980-1996*. New Brunswick, NJ: Rutgers University Press.

Horowitz, S., & Rosati, F. (2014, 4 September). 53 million Americans are freelancing, new survey finds. Freelancers broadcasting network.

Howe, N., & Strauss, W. (2009). *Millennials rising: The next great generation*. New York: Vintage Books.

Jenkins, H., Ford, S., & Green, J. (2015). *Cultura transmedia: la creación de contenido y valor en una cultura en red*. Barcelona: Editorial Gedisa.

John, N. A. (2012). Sharing and Web 2.0: The emergence of a keyword. *New Media and Society, 15*, 167–182.

John, N. A. (2013). The social logics of sharing. *Communication Review, 16*(3), 113-131.

Kamenetz, A. (2006). *Generation Debt: Why now is a terrible time to be young*. New York: Penguin Books.

Keen, A. (2016). *Internet no es la respuesta*. Barcelona: Barcelona.

Kennedy, J. (2015). Conceptual boundaries of sharing. *Information, Communication & Society, 19*(4), 461-474.

Kinder, T. (2002). Emerging e-commerce business models. *European Journal of Innovation Management, 5*(3), 130-151.

Kunz, W. M. (2007). *Culture conglomerates: Consolidation in the motion picture and television industries*. Lanham, MD: Rowman & Littlefield.

Lessig, L. (2006). *Code and other laws of cyberspace*. New York: Basic Books.

Lyubareva, I., Benghozi, P. J., & Fidele, T. (2014). Online business models in creative industries. *International Studies of Management and Organization, 44*(4), 43-62.

McAfee, A., Brynjolfsson, E., Davenport, T. H., Patil, D. J., & Barton, D. (2012). Big Data: The management revolution. *Harvard Business Review, 90*(10), 61-67.

McChesney, R. W. (2001). Global media, neoliberalism, and imperialism. *Monthly Review, 52*(10).

Mannheim, K. (1952). The *Problem of Generations' in Mannheim, K. Essays on the Sociology of Knowledge* (First Published 1923). London: RKP.

Mankiw N. G., Swagel P. L. (2006). The Politics and Economics of Offshore Outsourcing. *Journal of Monetary Economics, 53*(5), 1027-1056.

Martel, F. (2010). *Cultura mainstream*. Madrid: Taurus.

Mashey, J. R. (1997, October). Big Data and the next wave of infrastress. *Presentation at Computer Science Division Seminar*, University of California, Berkeley.

Mattelart, A. (1995). Excepción o especificidad cultural. *Telos: Cuadernos de Comunicación, Tecnología y Sociedad*, (42), 15-27.

Mattelart, A. & Vitalis, A. (2014). *De Orwell al cibercontrol*. Barcelona: Gedisa.

Mesenbourg, T. L. (2001). *Measuring the digital economy*. Washington, DC: U.S. Bureau of the Census.

Miller, T., Govil, N., McMurria, J., Maxwell, R., & Wang, T. (2005). *Global Hollywood 2*. London: BFI Publishing.

Morley, D. (2006) Globalisation and cultural imperialism reconsidered: Old questions in new guises. In J. Curran & D. Morley (Eds.), *Media and cultural theory* (pp. 30-43). New York: Routledge.

Naím, M. (2013). *El fin del poder*. Madrid: Debate.

Nye, J. S. (2011). *The future of power*. New York: PublicAffairs.

O'Reilly, T. (2005). What is Web 2.0? O'Reilly Network. Retrieved 23 February 2017.

Osterhammel, J., & Petersson, N. P. (2003). *Globalization: A short history*. Princeton, NJ: Princeton University Press.

Osterwalder, A., Pigneur, Y., & Tucci, C. L. (2005). Clarifying business models: Origins, present, and future of the concept. *Communications of the Association for Information Systems, 15*, 1-43.

Papacharissi, Z. (2009). *The virtual sphere 2.0: The Internet, the public sphere, and beyond*. In Routledge handbook of Internet politics. Chadwick, A., Howard, P. (Eds.) (pp. 230-245). New York: Routledge.

Pine, B. J. (1993). *Mass customization: The new frontier in business competition*. Boston: Harvard Business Press.

Pine, B. J., Victor, B., & Boynton, A. C. (1993). Making mass customization work. *Harvard Business Review, 71*(5), 108-111.

Ransdell, S., Kent, B., Gaillard-Kenney, S., & Long, J. (2011). Digital immigrants fare better than digital natives due to social reliance. *British Journal of Educational Technology, 42*(6), 931-938.

Rendueles, C. (2013). *Sociofobia*. Madrid: Capitan Swings.

Rheingold, H. (2000). *The virtual community: Homesteading on the electronic frontier*. Boston: MIT Press.

Rifkin, J. (2011). *La Tercera Revolución Industrial*. Barcelona: Capitan Swings.

Rifkin, J. (2014). *The zero marginal cost society*. New York: Macmillan.

Rudolph, C. W., & Zacher, H. (2016). Considering generations from a lifespan development perspective. *Work, Aging and Retirement, 3*(2), 113-129.

Rushkoff, D. (2010). *Program or be programmed: Ten commands for a digital age*. New York: Or Books.

Rushkoff, D. (2016). *Throwing rocks at the Google bus: How growth became the enemy of prosperity*. New York: Portfolio/Penguin.

Sackett, P. R. (2002). *Letter report*. Washington, DC: National Academies, Division of Behavioral and Social Sciences and Education.

Schiller, H. I. (1971). *Mass communications and American empire*. Boston: Beacon Press.

Schwab, K. (2017). *The Fourth Industrial Revolution*. New York: Crown Business.

Simpson, S., Puppis, M., & Van den Bulck, H. (Eds.). (2016). *European media policy for the twenty-first century: Assessing the past, setting agendas for the future*. London: Routledge.

Steel, P. Y., & Kammeyer-Mueller, J. (2015). The world is going to hell, the young no longer respect their elders, and other tricks of the mind. *Industrial and Organizational Psychology, 8*(3), 324-408.

Steemers, J. (2015). Broadcasting is dead. Long live television: Perspectives from Europe. In J. Trappel, J. Steemers & B. Thomass (Eds.), *European media in crisis: Values, risks and policies* (pp. 64–81). London: Routledge.

Stross, R. (2011). *The launch pad: Inside Y Combinator*. New York: Penguin.

Susskind, R., & Susskind, D. (2015). *The future of the professions: How technology will transform the work of human experts*. Oxford: Oxford University Press.

Sützl, W. (2014). The anti-economy of sharing. *Else Art Journal, 1*(0), 122–140.

Tapscott, D. (1997). *The digital economy: Promise and peril in the age of networked intelligence*. New York: McGraw-Hill Education.

Tapscott, D. (2009). *Grown up digital*. New York: McGraw Hill.

Tapscott, D. (2014). *The digital economy: Rethinking promise and peril in the age of networked intelligence* (Anniversary ed.). New York: McGraw-Hill Education.

Tapscott, D., & Williams, A. D. (2006). *Wikinomics: How mass collaboration changes everything*. New York: Portfolio.

Teece, D. J. (2010). Business models, business strategy and innovation. *Long Range Planning, 43*, 172-194.

Terranova, T. (2000). Free labor: Producing culture for the digital economy. *Social Text, 63, 18*(2), 33-58.

Toffler, A. (1980). *The third wave: The classic study of tomorrow*. New York: Bantam.

Tomlinson, J. (1991). *Cultural imperialism*. New York: Wiley.

Tseng, M. M., & Jiao, J. (2001). Mass customization. In G. Salvendy (Ed.), *Handbook of industrial engineering, technology and operations management* (3rd ed., pp. 684-709). New York: Wiley.

Tseng, M. M., & Piller, F. (Eds.). (2011). *The customer centric enterprise: Advances in mass customization and personalization*. Berlin: Springer Science & Business Media.

Tsui, K. K. (2016). Economic explanation: From sharecropping to the sharing economy. *Man and the Economy, 3*(1), 77-96.

Turner, G. (2015). *Re-inventing the media*. London: Routledge.

UNCTAD. (2013). *World investment report 2013: Global value chains investment and trade for development*. New York: United Nations.

Van Deursen, A. J., & Van Dijk, J. A. (2014). The digital divide shifts to differences in usage. *New Media and Society, 16*(3), 507-526.

Van Dijck, J., & Nieborg, D. (2009). Wikinomics and its discontents: A critical analysis of Web 2.0 business manifestos. *New Media and Society, 11*(5), 855-874.

Vogelstein, F. (2008). Weapon of mass disruption: The untold story of the making of the iPhone—and how Apple transformed the wireless industry forever. *Wired, 16*(2), 118.

Wasko, J. (2013). *Understanding Disney: The manufacture of fantasy*. New York: Wiley.

Weber, R. H. (2010). Internet of things: New security and privacy challenges. *Computer Law and Security Review, 26*(1), 23-30.

Weber, R. H., & Weber, R. (2010). *Internet of things* (vol. 12). New York: Springer.

Wei, J. (2014). How wearables intersect with the cloud and the Internet of things: Considerations for the developers of wearables. *IEEE Consumer Electronics Magazine, 3*(3), 53-56.

Wittel, A. (2011). Qualities of sharing and their transformations in the digital age. *International Review of Information Ethics, 15*(9), 3–8.

Womack, J. P., Jones, D. T., & Roos, D. (1991). *The machine that changed the world: The story of lean production*. New York: Harper Collins.

Wu, T. (2016). *The attention merchants*. New York: Knopf.

Xia, F., Yang, L. T., Wang, L., & Vinel, A. (2012). Internet of things. *International Journal of Communication Systems, 25*(9), 1101.

Zott, C., Amit, R., Massa, L. (2011). The business model: recent developments and future research. *Journal of Management, 37*(4), 1019-1042.

5 How important can content management be to a radio company? A case study of RTP and Rádio Renascença

Miriam Rodríguez-Pallares and
Paulo Faustino

5.1 Introduction

Radio professionalisation in Portugal started in the 1930s. Since then, this medium has been subjected to several challenges—not only social but also economic and technological ones—that acted as incentives for reconversion of the traditional radio business model.

After the failure to implement the European standard for digital radio—*digital audio broadcasting* (DAB)—the Portuguese radio model clung strongly to the distribution of online content (Correia, 2013). It was an indispensable qualitative leap in order to survive in the market. The use of the Internet as a broadcasting channel has brought in the multiplication of content in both number and format. Digitisation has blurred the actual characteristics of every traditional medium, including radio, in such a way that nowadays a multichannel, digital, asynchronous, fragmented and multi-format medium fosters bidirectional communication and hypertext possibilities (Cebrián, 2001; López, 2011; González, 2011).

However, the media industry as a whole has been forced to reformulate not only its functional model from a journalistic perspective—in terms of production and distribution—but also its corporate model (Faustino, 2009: 180)—that is, its organisational strategy. Thus, media concentration, globalisation of markets and the search for inverse marketing for online distribution channels have acted as catalysts for a new radio model that is dependent on a competitiveness. By virtue of competitiveness, the corporations—from this and other sectors—are searching for 'sectorial and individual improvements through investigation in order to foresee new ways of business or future higher profits' (Rodríguez-Pallares, 2012: 474). Adaptation of this new communicative and corporate paradigm requires adoption of organisational policies, particularly in the functional and strategic fields, that add value to original resources and position these corporations within their competitive field. Amidst these organisational policies, the priority is given to content management or the explicit representation of radio activity, which is to say that it is about managing and reinvigorating of the radio archive.

Content management is a complex process that involves confluence of 'technological components, corporate strategy and professional practices' (Fernández-Sande, Rodríguez-Barba and Rodríguez-Pallares, 2013: 393). Its objective is to treat and conserve data, structured and non-structured information and the explicit knowledge in the organisational digital context, which is susceptible of being used and reused in the execution of work processes and service provision to users (Eíto-Brun, 2013b: 378). Placing the content management theory into practice in a media communication corporate setting, one may come across an important peculiarity one needs to take into account: the production of a communication medium is intangible, and content management is a passive subject. This may presume a problem at the time of conceiving the media corporation solely as a productive unit; there is a risk that only the content connected to production (informative or for entertainment) will be subject to management. Therefore, the content generated in all other areas of corporate activity—such as HR, legal department, commercial department and department of technological surveillance—is kept in a decentralised way and ignored in the integrated process of describing, connecting, maintaining and broadcasting.

In that sense, bearing in mind the unique characteristics of the media industry, one understands that in this context content management policies divide skills in two areas of practice that correspond to two different but complementary perspectives: the strategic one and the functional one (Fernández-Sande et al., 2013: 393). Together, both respond to two objectives of the media sector: the general one, which is about obtaining economic benefits, and the specific one, which is about informing, educating and providing quality entertainment (Nieto and Iglesias, 2000):

- From the strategic perspective, which is connected with the purpose of obtaining long-term economic benefits (Casadesus and Ricart, 2010: 196), content management, as a process or a tactic determined by the business model, seeks to integrate and add value to all explicit resources generated in different areas of organisational activity in accordance with the corporate goals.
- From the functional or media perspective, which is connected to the process of content production and its storage, content management seeks to optimise the process of content production that, in relation to media and technological convergence, has multiplied its value as a newsworthy product, independent of the broadcasting channel.

Starting from these ideas, this study grounds itself on the following initial questions: Does the business model of Portuguese general broadcast companies contemplate content management as a piece of its organisational ecosystem? If the answer is affirmative, does the content add value as a corporate asset from the strategic and functional perspectives, or, on the contrary, do the policies implemented in the radio setting focus solely on storing content based on the old-fashioned and static idea of a watchdog?

5.2 Managing of radio content as a part of corporate strategy: The organisational integration

In their search for corporate dominance, corporations try to achieve competitive leverage through strategic policies or 'the combination of the company's offensive or defensive actions in order to position itself, affect and/or anticipate the market with the purpose of creating and developing a long-term competitive leverage' (Porter, 1999, quoted in Barceló, 2001: 38). These policies must go far beyond production factors—capital, labour and available resources—basing themselves, on the one hand, on technological and organisational innovation, which has an objective the optimisation of products and services, and, on the other hand, on leadership (Faustino, 2008: 53). It implies that a constant critical watch over corporate activity is necessary, both close and distant, and it requires allocating economic and human resources to it.

Strategic and innovative interest has generated a lot of investigative and management practices in the corporate field (Bouchikhi and Kimberly, 2001)—*technological surveillance*, *competitive intelligence* and *knowledge management* or *content management*, among others. These practices are based on the re-evaluation of a company's intangible assets, which may include the company's image, brand's value, management of knowledge, reputation and content. All these activities can be united under the same conceptual umbrella: *organisational intelligence* (OI), which is based on the corporate organisation as a whole and on single production units.

> [OI is the] effective use of the information in the field of organisation and of every internal operation, of knowledge—in its broadest sense— existing outside and inside the organisation, at the service of innovation or continuous improvement, a better use of opportunities, the creation of new knowledge and value, the education and well-being of people inside the organisation, as well as of the clients and other interested parties and society as a whole (Núñez, 2004).

According to Núñez, OI is considered to be a concept that integrates a wide array of applicable subsystems within a more specific corporate setting. In addition, from a theoretical perspective, content management includes production and expertise that are implicit in every line of business in a company. Silent expertise is a knowledge management skill (Rodríguez-Pallares, 2016)—it is about integrating all the knowledge generated by an organisation into one computer system (CMS, or *content management system*) in a centralised way. As an organisational unit, it achieves total control of its production and can reuse it in the most immediate way according to its general or particular interest, and in consonance with its business model.

In that sense, it justifies the strategic importance of content management in the context of OI as a co-helper in obtaining benefits not just in the short term but also in the long term, cooperating with organisational innovation and favoring making future decisions in a responsible manner. Nevertheless, the objective here is to identify from a strategic perspective whether or not there is a model of content management in the Portuguese radio setting that binds the company's production in every business area in a single database in a manner that, through its diffusion, allows cooperation between different organisational areas, facilitates innovation and develops long-term competitive leverage.

5.3 Content management in radio production: Media integration

If one restricts the field of analysis from the general (a company) to the particular (a radio company), implementation of content management focuses on media content or the content that can be mediatised. However, from a functional or media perspective, content management is responsible for radio production and its sources across their entire lifespan.

As previously stated, radio production and its sources acquire special relevance as a consequence of transferring from their sector to online distribution, characterised by an asynchrony that multiplies and gives special meaning to the content, since 'more than thinking about programming, one needs to think about contents that work individually, or in several different contexts, like archives, thematic segments, specialized channels or even in consonance with other languages and contents that are non-audio' (Correia, 2013: 167). Management of media content focuses its efforts on three points: optimising the production process, ensuring its storage and facilitating the immediate reuse of radio material with the objective of providing increased value to its media offerings.

Optimising the production process requires seeking integration in all processes common to content production and avoiding duplication of tasks in the production chain whenever media resources should be adapted to different distribution channels (García-Lastra, 2012: 172). The goal is producing once and broadcasting as many times and through as many channels as possible, which is known as POPE (produce once and publish everywhere). Rather than having a negative effect on on-air distribution, this approach complements it, as was demonstrated by the studies of new business models based on convergence (for example, Micó, Masip and Domingo, 2013; García-Avilés et al., 2014; Franklin, 2012)—even when the majority of scientific research was focused on print media or TV companies (Salaverría, 2015: 224). How can one achieve product economisation? The answer is centralisation of every resource explicit in a shared database, taking into account the criterion of informative transcendence—i.e., according to pre-established patterns at the time of selecting the content that responds to the news demand of each medium. The operability of content management depends, on this occasion, on two factors: on the one hand, the creation of an exhaustive working model and a cooperative work setting that connects documentarians and writers, taking into account the theories that defend the integration of the media production process or 'integrated journalism' (Salaverría, Negredo and Piqué, 2008); and, on the other hand, the use of a functional reach that, in the best of cases, has to encompass every radio production process identified with the same brand or brands that, in consequence of media integration, belong to the same mother company or to a multi-company group.

In order to secure control and facilitate the reuse of content, it is necessary to consider a computer platform, a CMS adapted to the needs of each organisational setting. This technological tool must provide a storage service to capture and store content, allowing editing and cooperative work and facilitating distribution of material following a pattern of user permissions in such a way that there is a hierarchy ensuring that the content base won't be manipulated.

> These content management processes must not be confused with business processes that constitute the chain of value in the organisation. ... The analysis of business processes is a previous step that must be made in order to identify the assets and the organisation's information needs (Eíto-Brun, 2013a: 46).

It is the exclusive right of the strategy and business model, not of content management, to decide which content and other intangible assets are to be subsumed within the radio database and how to give them organisational value. When identifying whether or not there is an integration of media content, that is when one will also be identifying what the business model is of a company group: that is, whether the organisational model is divided into brands and products—a common practice in large media conglomerates (De Mateo, Bergés and Sabater, 2009) that connect their brands from a productive perspective—or whether the brands and products are totally independent. And it is the prerogative of the content management policy to choose, within the model marked by business policies, content that observes criteria of informative transcendence, as well as to store and facilitate its reutilisation.

Thus, at the time of giving value to content as an intangible asset of a company, the policies of content management and the business model should be taken into consideration. One can then identify three phases at the time of giving value to this content: evaluating or identifying relevant content for the company regarding the organisational strategy and its business guidance, transforming information into knowledge by storing explicit knowledge in databases and maximising its reutilisation and creating value through a correct policy of content management in which there is not only storing and distribution but also content creation (Davenport and Prusak, 2001).

However, from a functional or media perspective, the purpose of this work is to identify what value is given to media content in the radio setting through the policies of content management and the business model, paying attention, on the one hand, to the process of selecting, storing, broadcasting and creating content by the departments responsible for management and, on the other hand, to whether or not the media content of a media group is interconnected in a single database.

5.4 Theoretical background

Content management is a concept that has gained relevance in the business field in the past few decades, a consequence, precisely, of this tendency for sectorial long-term growth through research, control and resource exploitation. Nonetheless, in case of radio, content management models seem to be developed according to a strategic perspective and are significantly different from one case to another on a functional level. One can identify two reasons to justify this situation:

- The first reason relates to the strategic perspective. It is explained through the previously mentioned peculiarity of media, whose production is itself dependent on management and claims all organisational attention, neglecting the integrated management of the company's production in other areas. It means that they are prioritising the functions of MAM (media asset management) and preventing the functions of CMS, whose objective is to manage transversal contents.
- The second reason relates to the functional or media perspective. It is connected to radio and its role in the media field, but above all, due to the uniqueness of its content in terms of exploitation, it is considered less profitable because image—the chief media product—is absent and because much content control is subordinated to authors' rights that limit its reutilisation in a free manner (article 91 of the *Code of Authors' Rights and Related Rights*).

For these reasons, media content management in the rádio field tends to be less favoured than TV content or digital channels' content.

In this sense, one can identify several examples of studies focused on the value of the media archive in TV or in the press, whether from a descriptive and practical perspective (Caldera, 2015; Rodríguez et al., 2012; Millard, 2003; Valenzuela and Echenagusía, 2008; Naseiro, 2013; Ariza, 2004; Villar, 2005) or related to storing (Marçal-Ferreira, 2013) or even taking into account the limits and good practices of reutilisation of information (Hidalgo and López, 2014). Although radio looks like a less attractive sector from the point of view of a scientific study in this field (perhaps, as stated above, due to its smaller potential for value added to its content), there have also been studies connected to archives and the recovery of material in particular cases (Nuño et al., 2007; Kischinhevsky, 2014; Marta and Ortiz, 2013).

Notwithstanding, a study carried out in the Spanish context (Rodríguez-Pallares, 2014), which can be emulated in the Portuguese context, can be used as direct precedent. The research carried out between July 2010 and April 2013 reviewed content management models in Spain's four most influential radio brands associated with EGM (Estudio General de Medios): Cadena SER, Onda Cero, COPE and RNE.

Conclusions of this study resulted in the categorisation of very disparate models from the case studies at SER, Onda Cero, COPE and RNE. These models are a consequence of the organisational and structural differences in each corporate field. However, each case also had a common factor: the policies of content management paid attention only to media or newsworthy materials; in no case were they identified as policies of integrated content management that unified in the same database the content generated by other business areas that are not directly connected to radio production (HR, legal, commercial, general management). Therefore, the objective of transversality and integration of organisational materials was neglected, leading one to advocate the strategic perspective over other perspectives. Content management in the Spanish radio sector is an under-researched area.

Therefore, with this contextual theoretical framework as a guide, we now examine the Portuguese radio field in accordance to the same descriptive categories.

5.5 Methodology, objectives and limits

In order to carry out this study, a sample of two general radio companies headquartered in Portugal was selected. The selection was based on audience rating indexes released by Grupo Marktest on April 2015. The selection was also based on the premise that OI theories are applicable not only to the private sector but also to the public one, which, nowadays, without the protection of unconditional governmental support, must pay attention to business outcomes, defining its profit objectives to avoid bankruptcy, and, at the same time, keep sight of its social responsibilities (Faustino, 2008). As a result, the sample selected is Rádio e Televisão de Portugal (RTP), a state-owned radio company, and Rádio Renascença, the general radio company with the highest nationwide audience ratings.

As stated before, the main purposes of this study are the following:

- O1—Strategic perspective: identifying the existence of an integrated and transversal content management model in different business areas of a company.
- O2—Functional or media perspective: identifying the existence of a content management model that integrates media content of the same mother company in the same database.

To achieve these purposes, the analysed reality has been segmented into different levels of analysis that serve as research phases in order to simplify the understanding and create a logical structure of the content.

- L1—Descriptive and organisational level: contextualising and identifying content management and its organisational positioning.

  The organic location of the department responsible for content management is understood as justifying, on the one hand, the value attributed to this activity on an organisational level and, on the other, the functional reach of the service. Therefore, presenting a general vision of the organisational corporate model and of the hierarchical positioning of the content management department helps to infer its responsibilities in the company or company group's activity to which it belongs, as well as its solely media-related responsibilities (media integration) and also its responsibilities that are common to all company resources (company integration).

- L2—Functional level: identifying the policies of content management from functional and strategic perspectives.

  Far from pretending to create a detailed description of the documental activity of the content management and archive departments, this study offers a brief approach reinforced by diagrams and tables regarding the most prominent traits of the work model used in the radio field. Thus, particular attention is paid to the following variables: documental process, technological support and recovery and reutilisation of contents. The approach to these variables contributed to assessing the importance of the content management department.

The research methodology is based on a multiple-case-study model or the systematic investigation of different realities based on the same model of analysis. In this context, the study is sustained mostly by qualitative research techniques (Table 5.1) whose epistemological basis, according to Stake, is 'existential (not deterministic) and constructivist' and emphasises interpretation (Stake, 2007: 46). Thus, starting from the collated data, one arrives at a series of interpretative conclusions that provide answers to the proposed objectives.

There are certain methodological limits of this research in sampling and temporal terms: the study focuses solely on the case of two general radio stations in the Portuguese context, and the analysis comprises only the period between April and June 2015. However, the results obtained are considered to be revealing enough to signify a symptomatic representation of reality in this media industry.

5.6 Case Studies of Portuguese Radio Archives

Content management in Portugal is regulated by article 83 of Law 54/2010, or the *Radio Law*. This article states that 'radio operators in a national and regional context must organize their audio and musical archives with the

Table 5.1 Summary of the study's methodological process

Investigation Phase	Scientific Methods	Considered Data
Selection of sample for analysis	Review of audience index rates (Breme Rádio study from Grupo Marktest)	• Audience share %
Theoretical framework precedents	Bibliographic review	• Scientific backup in the field of intangible assets management in the company's surroundings, radio content management, radio and legal context in Portugal • Literary backup: corporate and historical context of the analysed cases
L1	In-depth interviews *	• The company's organisational structure • Organisational positioning and the department's organic dependence
L2	- In-depth interviews * - Nonparticipating direct observation **	• Documental process • Technological tools • Access and material recovery models • Integration of content from different business areas • Integration of media content from the analogue and online areas • Identification of the documentalist as the author of on-air and online contents

* Conducted between April and June 2015, following the criterion of organisational responsibility in the settings of the analysed content management.
** Procedure carried out in conjunction with the in-depth interviews.

purpose of conserving the registers of public interest' and adds with nuances, increasing the legal obligation, that in the case of radio, public-service-specific obligations include, among other aspects, 'keeping and updating audio archives, insuring the maintenance, updating and availability to the public, according to museum principles and rules applicable, of a collection representative of the evolution of the radio medium in terms of a licensing contract' (Law 54/2010). Regarding the period considered, the aforementioned law clarifies in article 39 that 'broadcasts must be recorded and kept for a minimum period of 30 days if a no more extensive one has not been determined by law or by judicial decision'.

Starting from national legal directives, one understands that the recording and storing of continuing radio broadcasts during the period of a month is grounded on a probative principle in order to avoid legal conflicts, and from these directives, one infers that it is the responsibility of each radio company to store its own content—except those subjected to authors' rights and related rights whose storing is stipulated by law. In other words, decisions on the design and implementation of a content management model are connected to the business model and the organisational strategy of each radio company, except in the public case, in which the company, by legal imperative, is obliged to provide value to its catalogues through correct storing and facilitate its access as historical audiovisual heritage. This is how it was stated in the document *New Options for the Audiovisual*, dated from 2002 and authored by the Ministry of Presidency.

Therefore, the lack of specific legislation regarding selection and preservation of audiovisual content in the radio field becomes evident. The existing legislation is generic and inadequate in practice to the reality of the organisations that adopt, as a single legal source, the *Code of Authors' Rights and Related Rights*, whose normative guidelines, on occasion, are not adjusted to the context of the institutions' functioning (Epifânio, 2013).

5.6.1 *Rádio e Televisão de Portugal (RTP)*

5.6.1.1 *Descriptive and organisational level*

In terms of corporate macrostructure, RTP is a media conglomerate that encompasses television and radio services in the Portuguese public sphere, an S.A. whose goal is to provide radio and television services regarding Portuguese context. Until 2004, public television (RTP) and radio (RDP) were independent entities; it was at this point that, after a process of media reorganisation under state guidance, both entities started a process of fusion with the creation of the holding company Rádio e Televisão de Portugal, SGPS, S.A., licensed to provide the public service of Portuguese radio. In 2007, following a process of restructuring, the organic public service entity began denominating itself Rádio e Televisão de Portugal S.A. in accordance with Law 8/2007 (February 14), subsequently changed in 2011 and 2014.

The RTP public service is, in turn, regulated in accordance with the Contract of Public Service Television Licensing of March 25, 2008, and the Contract of Public Service Radio Licensing of June 20, 1999. The public service status is precisely what defines the storage and archive management strategy, no matter whether the company is TV- or radio-related—in that the probative meaning of the materials acquires an important role as a reflection of the history of Portuguese public service radio.

As a logical consequence of the changes affecting RTP, the department responsible for content management has also undergone modifications. Thus, after the 2004 fusion and the 2007 restructuring, television and radio archives were unified under a single organic direction, the Broadcast and Archive Direction, staffed with 55 workers. It implies that, in organisational terms, the RTP archives are understood in a generic way as audiovisual archives, even when, in management terms, there are independent work teams and tools for each documental typology.

The coordinating organ of content management activity depends on the Institutional Relations and Archives area, which, in turn, directly depends on the Board of Administration. From this position, it is possible to recognise organisational attributes to content management; however, as mentioned before, it belongs to the Broadcast and Archive Direction, which implies a direct connection of the archive only with media production, neglecting the content of other areas of activity in the company and fomenting the decentralisation of organisational content.

In fact, the Portuguese public radio archive is concerned only with media or newsworthy content, which can be seen in the division of its activity into two collections. On the one hand, the historical archive, founded at the same time as the Emissão Nacional (mother company of Portuguese public radio, later RDP and then RTP), includes radio-produced programmes of information, entertainment and music, and on the other hand, the musical archive, created in a permanent manner in 1972 but fully operational only after the Carnation Revolution—25 April 1974—includes music collections and commercial records that have been used in the past 80 years, as well as some editions and productions of Portuguese radio.

5.6.1.2 *Functional level*

The content management model in RTP is founded on three aspects: documental process, technological support and recovery and reutilisation of content (Table 5.2).

The media documentation process is applied to a selection of content that forms the permanent archive, provided that authors' and related rights have expired. It means that here RTP definitely keeps all radio content of its own authorship. If the content has been acquired or is subject to third parties, RTP keeps the content while the law permits the company to use it. The process varies according to the typology of the content and the cessation contract. The moment the content can no longer be used without paying new royalties, it is eliminated.

Bearing this in mind, RTP keeps only the broadcasted content in the permanent archive, except for non-edited material (contrary to what happens with television), and does it in two versions, high- and low-quality (browser). There are twin copies that are connected. Both the preservation masters (in BWF or *broadcast wave format*) and the access copies (in MP2) are recorded on linear tape open (LTO).

Once the content has been selected, it is described according to normalised documental languages of RTP's own creation—although based on international recommendations by organs like the European Union, the International Archives Council or the ISAD—and integrated into databases

Table 5.2 Functional summary of Rádio Pública Portuguesa (RTP)

Documental process	
Policies of selection and custody	Conditioned by compliance with authors' rights and related rights: only content owned by RTP is kept definitively; if that is not the case, it is kept only while it can be used, after which point it is eliminated. Only the on-air broadcast is stored; non-edited material is discarded except in cases of special programs or content.
Controlled languages	Thesaurus of in-house creation.
Typology of contents	Audio.
Digitalization of materials	Yes, it's in process; however, they are not integrated with the databases. For now, digitised content is referenced in an Excel file, in order for pertinent changes to be carried out later on and for descriptors to be added in the current databases.
Technological tools	
Content management system	No, they rely on a database of in-house creation that contains registers, not content.
Implementation	After RTP and RDP's fusion and the creation of the archive under the Broadcast and Archive Directorate.
System integrated with the editorial office's software	No, the archive is accessed through the intranet, and the input is not connected to the editorial office's department.
Access and recovery	
Manuel or automatic recovery	Manual.
Internal access profiles for content	There are access profiles, operated by username and password.
Outgoing channels	As a public service, RTP releases its content independently of the solicitor and in accordance with a price table for public access (http://www.rtp.pt/wportal/grupo/contactos_pdf.php).
Security and control of the materials	
Security system	Copies of the archives are kept, stored in buildings adjacent to RTP's own facilities.
Legal copy	Since 2009, storage has become permanent; after the three months of legal obligation, the content is digitally kept apart from the archive. This content is used to satisfy requests from citizens.
Integration	
Integration of media material from the analogue and online business line	No.
Integration of media material from other work areas	No.
Adding value to content management	
Creation of content by those responsible for the department	Yes, for the web (http://www.rtp.pt/arquivo/). There is a person responsible for the creation of collections who relies on the cooperation of the TV and radio archive team.

of its own creation, developed in the 1990s. In these databases, the content is referenced but not included. What does that mean? It means that the recovery of material is manual, not automatic. Although at the moment the radio archive is being digitised, it is not integrated into a digital robotised archive.

Concerning the computer tool supporting management activity, it is necessary to realise the disparity on the technological and organisational levels between the models of television and radio management. Thus, while the television managing system relies on an advanced tool based on AVID technological performance, the radio managing system is far more rudimentary. It is backed up by simple documental databases of RTP's own devising whose implementation dates from the 1990s, even though the company plans to jump to the digital model with the goal of automatising the whole process of radio material intake and recovery.

The following points must be considered in regard to the processes of integration:

- Media integration: At the moment, there are only two functioning divisions or areas of activity, television content and radio content, that are managed independently and supported by computer systems and independent working teams. The treatment of the latter, compared to that of the former, indicates it is clearly undervalued. At the time of this writing, the online contents are not managed:

    We are not, at the moment, creating Internet archives. It's also true that, until now, save a few exceptions, there are no Web-exclusive contents; they're built with basis on TV and radio; once we have exclusive contents, then, yes, we'll have to start doing it. … This is a matter that we'll have on the table because there will be unique contents for the online field. (Hilário Lopes, personal communication, 21 April 2015)

- Organisational integration: The archive of RTP's materials curates only media or newsworthy content in TV terms or in radio terms; there is no policy connected to centralisation of transversal content from different areas of the state-owned media group.

5.6.2 *Rádio Renascença (RR)*

5.6.2.1 *Descriptive and organisational level*

Rádio Renascença is a Portuguese radio brand whose programming focuses on information and entertainment content. It is managed by the Catholic Church through the Lisbon Patriarchy and belongs to the communication group Renascença Comunicação Multimedia R/Com, at the time the manager of the radio chains RFM, Mega Hits, Rádio Sim and Intervoz, which, although directed at different targets, operate under the precept of Catholic ideas.

The history of Rádio Renascença is strongly linked to the nation's history: it was one of the first nationwide broadcast companies; it originated in the 1930s and has fit into a context of radio popularisation that served as a stimulus to seek new evangelisation strategies for the Church. This network is known for its role in the events that marked the Portuguese history on 25 April 1974, since it provided a broadcast channel from which the revolutionary forces could read their dispatches. Afterward, even when it was occupied in 1975 and lived through many internal changes—fighting for the structure and the basis of its identity (Ribeiro, 2001)—it was the only radio company not nationalised during the revolutionary period. It managed to be excluded from the 2 December 1975 law, which set out the restructuring and nationalisation of radio networks of wider reach in Portugal.

Currently, Rádio Renascença, according to data from the Breme Rádio study, is the ratings leader due to an uninterrupted increase of listeners that skyrocketed in the last two decades of the 20th century. Although regular broadcasts started in 1937, it was in 1990 that Rádio Renascença created the Center of Documentation, which, although having undergone several changes related to its scope and performance, currently stands as a stable department, depending directly on the management of the radio company. At one point, it was staffed with four workers, but at the moment, only one person runs the department. What does this mean? It means that the organisational positioning is adequate according to the precepts marked as guides, although the department's dimension limits its functions.

5.6.2.2 *Functional level*

Like in the previous case, the pillars of the content management and archival model are divided into three points: documental process, technological support and recovery and reutilisation of contents (Table 5.3).

Regarding the documental process, Rádio Renascença also limits its activity to media or newsworthy content. Notwithstanding this, in this context, and unlike RTP, the broadcast continuum of the radio is kept only for a month, in accordance with the law. The permanent archive is composed of a selection of material that professionals who have decision capacity, who use a criterion for informative transcendence and who are editorially responsible within the company recommend to the Center of Documentation. All the audio selected by the editorially responsible people is transferred to the Center of Documentation in wave format and with a short description—name, date, programme and generic theme. It is at the Center of Documentation that data are transformed into MP3 format, analysed and their content described in detail, indexed and kept in the pertinent database.

The materials comprising the archive are of textual and audio typology. To manage them, the Center of Documentation has two software programmes: a computer programme developed internally in 2013 that allows storage of documents in PDF and MP3 and that is where digital content

Table 5.3 Functional summary of Rádio Renascença

Documental process	
Policies of selection and custody	They're conditioned by compliance with authors' rights and related rights: they only keep definitely the contents selected by editorial office chiefs; the broadcast is kept for one month.
Controlled languages	European Union's EuroVoc Thesaurus. Adapted UDC.
Typology of contents	Audio and text.
Digitalization of materials	In process. It started in 2002 and is being developed with the company's in-house technical and human resources. At the moment, the paper format and the cassette tapes and DAT are being digitised.
Technological tools	
Content management system	There are two software programs: an in-house one for the press and audio archive, and another one acquired from the Doc Master company that works as a bibliographic and musical catalogue.
Implementation	Doc Master (2002), Internal tool (2013).
System integrated with the editorial office's software	No, there are two distinct tools with different formats. The archive is accessed through the intranet, and the input is not connected with the editorial office's software.
Access and recovery	
Manual or automatic recovery	Automatic.
Internal access profiles for content	The access is universal to R/Com; downloading is restricted by username and password.
Outgoing channels	There is no regulation regarding it; however, some content has been commercialised or facilitated for special occasions.
Security and control of the materials	
Security system	Through internal backup, managed by the computer staff.
Legal copy	One month.
Integration	
Integration of media material from the analogue and online business line	No.
Integration of media material from other work areas	No.
Adding value to content management	
Creation of content by those responsible for the department	No.

taken from the national press and the audio archive is stored and an older bibliographic and musical database, created in 2002 by an external company (Doc Master) according to the center's needs, that works as a bibliographic and musical catalogue—that is, as a keeper of the historical disc collections and the monographs.

Access and reutilisation of the archive's materials are connected to the nature of functions that each worker carries out within the company: all the workers at R/Com have the right to access the archive's contents online; however, only those who have responsibilities for the network's content can import data to their workstations.

Regarding the process of integration, in this case, it is also necessary to clarify the following:

* Media integration: Rádio Renascença keeps in a centralized way the content selected by the responsible staff members, as well as the historical disc collection—constituted especially of vinyl works that date from the beginnings of the Rádio Renascença—and textual content. However, each network in the R/Com group manages its own disc collection; i.e., the musical content is decentralised. Conversely, although Rádio Renascença has invested in integrating the online and conventional editorial offices, the databases are not shared; there is a database *ex professo* designed to keep website materials, resulting from the specificity of multimedia content, and on another level, there is the Center of Documentation.
* Organisational integration: The content from different areas of business is not centralised in the same database. 'The archive of the Center of Documentation is constituted of the content produced by the radios and a combination of references, newspaper articles and musical items that serve as a support to the content broadcast by the radio brands' (Ana Isabel Almeida, personal communication, 10 July 2015).

5.7 Concluding reflections

The content management models analysed present notable structural, organisational and functional differences. However, the results suggest that only media or newsworthy content is managed and that in no case were policies of transversal integration identified from either an organisational (strategic) perspective or a media-related (functional) perspective that allow optimisation of content from the OI perspective. This way, one does not see either company making full use of the possibilities of a unique CMS, a database of different documental typologies, which implies a duplication of work and economic resources.

Although there are models of content management that give more importance either to the heritage value of the materials (RTP) or to the usability and practicality of the same (Rádio Renascença), in both cases one detects a clear under-exploitation of the potential of content management for the company's strategy regarding the integration and reinvigoration of the company and media content and, in contrast, the creation of unique content.

In this sense, the data obtained in the Spanish and Portuguese studies coincide, so that one could claim that the practice of content management at radio companies of wider reach deals with functional and strategic limitations and, above all, lacks a normative framework that could standardise management of documental materials.

Finally, it seems important to point out that substantial changes in the models of content management under analysis are not contemplated in the short term, even though RTP has in mind implementation of the logic of the TV business in radio in terms of documental management, digitalising its archive and improving its radio computer support.

References

Ariza, R. M. (2004). El archivo de la palabra de radio nacional de España. *Revista General de Información y Documentación, 14*(2): 29-58.

Barceló, M. (2001). *Hacia una economía del conocimiento*. Madrid: ESIC.

Bouchikhi, H., and Kimberly, J. R. (2001). It's difficult to innovate: The death of the tenured professor and the birth of the knowledge entrepreneur. *Human Relations, 54*(1): 77-84. DOI: 10.1177/0018726701541010

Caldera, J. (2015). El principio de Pareto en el control documental de programas informativos televisivivos: Implicaciones en el Media Asset Management. *Revista Digital de Biblioteconomia e Ciência da Informação, 13*(3): 480-490. DOI: 10.20396/rdbci.v13i3.8639461

Casadesus, R., and Ricart, J. (2010). From strategy to business models and on to tactics. *Long Range Planning, 43*: 195-215. Available at http://www.businessmodelcommunity.com/fs/Root/8oex1-Casadesus_et_Ricart.pdf (accessed 3 August 2015). DOI: 10.1016/j.lrp.2010.01.004

Cebrián, M. (2001). *La radio en la convergencia multimedia*. Barcelona: Gedisa.

Correia, S. (2013). Do insucesso do DAB à expansão online: A estratégia digital da rádio pública portuguesa. *Observatrorio (OBS*) Journal, 7*(2): 161-181.

Davenport, T. H., and Prusak, L. (2001). *Conocimiento en acción: Cómo las organizaciones manejan lo que saben*. Buenos Aires: Pearson Education.

De Mateo, R., Bergés, L, and Sabater, M. (2009). *Gestión de empresas de comunicación*. Sevilla-Zamora: Comunicación Social, Ediciones y Publicaciones.

Eíto-Brun, R. (2013a). *Gestión de Conetnidos*. Barcelona: UOC.

Eíto-Brun, R. (2013b). Madurez de la gestión de contenidos, ¿sinónimo de desgaste o de oportunidades? *El Profesional de la Información, 22*(5): 377-380. DOI: http://dx.doi.org/10.3145/epi.2013.sep.01

Epifânio, N. M. (2013). Enquadramento legal dos arquitos audiovisuais e sonoros em Portugal. *Informação & Informação, 18*(3): 222-237. DOI: 10.5433/1981-8920.2013v18n3p222

Faustino, P. (2008). Restructuring and turnaround of a public service broadcaster: Public management with private attitude. *Journal of Media Business Studies, 5*(3): 53-81. DOI: 10.1080/16522354.2008.11073475

Faustino, P. (2009). Tendências e dinâmicas do mercado dos media em Portugal. *Comunicação e Sociedade, 16*: 177-212. DOI: 10.17231/comsoc.16(2009).1037

Fernández-Sande, M., Rodríguez-Barba, D., and Rodríguez-Pallares, M. (2013). La gestión de contenidos como actividad estratégica en empresas de radiodifusión: Estudio de casos en la radio comercial española. *El Profesional de la Información, 22*(5): 392-398. DOI: 10.3145/epi.2013.sep.03

Franklin, B. (2012). The future of journalism: Developments and debates. *Journalism Studies, 13*(5): 663. Available at http://www.tandfonline.com/doi/full/10.1080/14616 70X.2012.712301 (accessed 2 February 2017). DOI:10.1080/1461670X.2012.712301

García-Lastra, J. M. (2012). Del modelo productivo de la era analógica al de la radio multicanal. In J. I. Gallego Pérez and M. T. García Leiva (Eds.), *Sintonizando el futuro: Radio y producción sonora en el siglo XXI* (pp. 167-188). Madrid: Instituto RTVE.

García-Avilés, J. A., Kaltenbrunner, A., and Meier, K. (2014). Media convergence revisited: Lessons learned on newsroom integration in Austria, Germany and Spain. *Journalism Practice, 8*(5): 573-584. DOI:10.1080/17512786.2014.885678

González, P. (2011). La radio en Internet: Las webs de las cadenas analógicas tradicionales. In M. A. Ortiz and N. López (Eds.), *Radio 3.0: Una nueva radio para una nueva era* (pp. 89-122). Madrid: Fragua.

Grupo Marktest. (2015a). Rádio no século XXI é coisa do futuro. Available at http://www.marktest.com/wap/a/n/id~18b8.aspx (accessed 10 July 2015).

Grupo Marktest. (2015b). *Vaga de abril 2015 do Bréme Rádio.* Available at http://www.marktest.com/wap/a/n/id~1ef4.aspx (accessed 10 July 2015).

Hidalgo, P., and López, I. (2014). Reutilización de imágenes de archivo en televisión: Derechos de propiedad y de uso. *El Profesional de la Información, 23*(1): 65-71. DOI:10.3145/epi.2014.ene.08

Kischinhevsky, M. (2014). Compartilhar, etiquetar: Interações no radio social. *Comunicação, Midia e Consumo, 11*(30): 143-162. DOI: 10.18568/1983-7070.1130143-162

López, N. (2011). La radio se transforma: Nuevas tecnologías, nuevos hábitos y nuevos perfiles para el medio más cercano. In M. A. Ortiz and N. López (Eds.), *Radio 3.0: Una nueva radio para una nueva era* (pp. 15-40). Madrid: Fragua.

Marçal-Ferreira, S. P. (2013). *O arquivo da rádio da RTP: Preservação do seu acervo.* Master's diss. Portugal: Universidade Nova de Lisboa.

Marta, C., and Ortiz, M. A. (2013). Gestión de los fondos documentales en Radio Nacional de España. *El Profesional de la Información, 22*(5): 399-403. DOI: 10.3145/epi.2013.sep.04

Micó, J. L., Masip, P., and Domingo, D. (2013). To wish impossible things: Convergence as a process of diffusion of innovations in an actor-network. *International Communication Gazette, 75*(1): 118-137. DOI:10.1177/1748048512461765

Millard, D. (2003). CBC archives. *Reference Reviews, 17*(3): 60-61. DOI:10.1108/09504120310466973

Naseiro, A. (2013). El archivo del diario 'Pueblo': Un referente para la historia de la prensa en España durante el franquismo y la transición democrática (The archive of the daily 'Pueblo': A reference to the history of the press in Spain for the Franco era and the Spanish transition). *Documentación de las Ciencias de la Información, 36*: 11-29.

Nguyen, G. D., Dejean, S., and Moreau, F. (2014). On the complementarity between online and offline music consumption: The case of free streaming. *Journal of Cultural Economics, 38*(4): 315-330. DOI:10.1007/s10824-013-9208-8

Nieto, A., and Iglesias, F. (2000). *La empresa informativa*. Barcelona: Ariel.

Núñez, I. A. (2004). La gestión de la información, el conocimiento, la inteligencia y el aprendizaje organizacional desde una perspectiva socio-psicológica. *Acimed: Revista Cubana de los Profesionales de la Información y la Comunicación en Salud, 12*(3). Available at http://scielo.sld.cu/scielo.php?script=sci_arttext&pid=S1024-94352004000300004 (accessed 25 July 2015).

Nuño, M. V., Sánchez, M. I., and Afuera, A. (2007). *Documentación en el medio radiofónico: Hacia un entorno digital*. Madrid: Síntesis.

Portugal, Assembleia da República. (1985). *Decreto-lei 63/85 (Código do direito de autor e dos directos conexos)*. Available at https://www.spautores.pt/assets_live/165/codigododireitodeautorcdadclei162008.pdf (accessed 15 July 2016).

Portugal, Assembleia da República. (2004). *Lei 50 de 24 de agosto*. Diário da República, Lisboa, n. 199, Série A. Available at https://dre.pt/application/dir/pdf1s/2004/08/199A00/56585665.pdf (accessed 15 July 2016).

Portugal, Assembleia da República. (2010). *Lei 54/2010 de 24 de dezembro (Lei da Rádio)*. Diário da República, Lisboa, n. 248, Série 1ª. Available at https://dre.pt/application/dir/pdf1s/2010/12/24800/0590305918.pdf (accessed 15 July 2016).

Portugal, Ministério de Comunicação Social. (1975). *Decreto Lei n° 674C/75 de 2 de Dezembro*. Available at http://www.gmcs.pt/ficheiros/pt/decreto-lei-n-674-c75-de-2-de-dezembro.pdf (accessed 15 July 2016).

Presidencia do Conselho de Ministros. (2002). *Novas opções para o audiovisual*. Available at http://www.gmcs.pt/ficheiros/pt/novas-opcoes-para-o-audiovisual.pdf (accessed 15 July 2016).

Rádio e Televisão de Portugal. (2015). *Governo da sociedade*. Available at http://media.rtp.pt/institucional/rtp/ (accessed 10 July 2016).

Rádio Renascença. (2015). *Ficha técnica*. Available at http://rr.sapo.pt/FichaTecnica.aspx (accessed 10 July 2016).

Ribeiro, N. (2001). Momentos marcantes na história da Rádio Renascença (1937–1987). In Para a história da rádio em Portugal. *Revista Observatório da Comunicação (4)*: 97-112.

Rodríguez, D., Hernández, T., and Méndez, E. (2012). Archivos y centros de documentación en la prensa local de la comunidad de Madrid (Local media archives in the region of Madrid). *Documentación de las Ciencias de la Información, 35*: 11-30. DOI: 10.5209/rev_DCIN.2012.v35.40444

Rodríguez-Pallares, M. (2012). El capital intangible como clave estratégica en la empresa radiofónica española: El caso de la Cadena SER. In M. Fernández-Sande and A. Adami (Eds.), *Panorâmica de Comunicação e dos Meios no Brasil e Espanha/Panorámica de la Comunicación y los Medios en Brasil y España* (pp. 473-495). São Paulo: Intercom.

Rodríguez-Pallares, M. (2014). *Análisis de los actuales modelos de gestión de contenidos y conocimiento en las grandes cadenas de radiodifusión españolas: SER, Onda Cero, COPE y RNE*. PhD thesis, Universidad Complutense de Madrid, Spain.

Rodríguez-Pallares, M. (2016). Propuesta conceptual de un modelo de gestión de contenidos y del conocimiento para la empresa radiofónica española. *Revista Española de Documentación Científica, 39*(2). DOI: 10.3989/redc.2016.2.1271

Salaverría, R. (2015). Ideas para renovar la investigación sobre medios digitales. *El Profesional de la Información, 24*(3): 223. DOI:10.3145/epi.2015.may.01

Salaverría, R., Negredo, S., and Piqué, A. M. (2008). *Periodismo integrado: Convergencia de medios y reorganización de redacciones*. Barcelona: Sol90.

Stake, R. (2007). *Investigación con estudio de casos*. Madrid: Morata.
Valenzuela, M. C., and Echenagusía, J. (2008). La gestión documental audiovisual de los archivos de televisión, estudio de caso en América Latina. *Biblios: Revista Electrónica de Bibliotecología, Archivología y Museología, 32.* Available at https://goo.gl/7mEmqY (accessed 10 July 2015).
Villar, D. (2005). Hacia una televisión pública 2.0: El Creative Archive de la BBC. *Comunicar* (25). Available at http://www.revistacomunicar.com/index.php?contenido=detalles&numero=25&articulo=25-2005-136 (accessed 15 July 2015).

Personal sources

Almeida, A. I., responsible for the RR Center of Documentation (10-07-2015).
Lopes, H., director of RTP Broadcast and Archives (21-04-2015).
Palmeiro, J., President of the Portuguese Press Association (16-04-2015).
Ramos-Pinheiro, J. L., RR management (16-06-2015).

6 Different uses of marketing tools Facebook and Twitter by Greek politicians during elections

Vagia Mochla and George Tsourvakas

6.1 Introduction

Nowadays, social media are believed to have an effect on political discourse and communication in our society. Specifically, social media are used increasingly in politics, constituting a marketing window, a surveillance tool and a medium of political implementation. Recently, micro-blog services like Twitter and social networking sites like Facebook demonstrated that they can contribute to increased participation in politics.

Undoubtedly social media have invaded our lives and constitute an integral part of it. More and more politicians, feeling the need to exploit the privileged space that social media offer so as to publish their personal messages, participate in creating a mosaic of information and experiences with fellow users. As social media have recently emerged as a platform for the exchange of social, informative and political considerations, it should not be surprising that politicians use these means of communication and marketing to influence public attitudes in their favour, to form agendas and even to shape the outcome of their political campaigns. With their continuing emergence as a means of marketing, disseminating information and political contradictions supporting political debate political contradictions, politicians are no longer responsible just for what they post but also for the resulting reactions, interactivity and dialogues they provoke with their potential voters. The interactivity via new marketing tools is very important today for politicians.

Therefore, this chapter explores the uses of the two main Internet marketing tools, Facebook and Twitter, in order to see how social media consolidate the diffusion of the political messages and the substantial work that the politician does during the election period. In the first section, there is a review of relevant literature on digital marketing tools. Based on this, three main research questions are developed, following the methodology of content analysis. The second section reveals the main results that answer the research questions. Finally, there is a discussion of the results in relation to the theoretical and practical issues. Limitations and future research conclude the study.

6.2 Conceptual Framework

6.2.1 *Social media as a marketing tool*

The use of social media in election periods in order to participate is an emerging trend, and their use for the purpose of information dissemination is widespread. Used as a marketing tool, social media allow users to communicate and provide an opportunity for political players through them to influence the perceptions of voters in favour of their own views and against those of their political opponents. It has been observed that in a very short time, politicians in modern democracies around the world have strongly adopted social media to engage their constituents in the election process and activate dialogue with them (Hong and Nadler, 2011).

The field of political marketing has been developing rapidly and has been chronicled by researchers (Shama, 1976, Lock, and Harris 1996, Harris, Lock and O'Shaughnessy, 1999; Dermody and Scullion, 2001; Baines, Harris and Lewis 2002, Hughes and Dann, 2009, Ormond, 2012). In 2009 (p. 244), Hughes and Dann defined the concept of political marketing as 'a set of activities, processes or political institutions used by political organizations, candidates and individuals to create, communicate, deliver and exchange promises of value with voter-consumers, political party stakeholders and society at large'. Additionally, the use of social media has been associated with a series of positive attitudes and certain behaviours on the part of the voters, including increased participation, especially of the young people (Bakker and de Vreese, 2011; Baumgartner and Morris, 2010); increased confidence and belief that they can better understand and influence political developments (Towner and Dulio, 2011); and increased support of the candidacy of a particular politician (Gilmore, 2012). Related studies on the use of social media have focused largely on the key results arising from the behaviour of users, including political participation and interaction with politicians, paying attention to the basic processes that lead to these results. However, there is a growing interest in evaluating the reliability of social media (Powell, Richmond and Williams, 2011).

The applications of social media are increasingly recognised by politicians and political parties as important tools (Bryer et al., 2010) to both develop relations with voters (Baumgartner and Morris, 2010; Gulati and Williams, 2010) and achieve an 'opening' to potential voters during election campaigns (Dean and Croft, 2001; Zhang et al., 2010). Nowadays, a growing trend is for research focusing on the role of social media in the political process, with the recent presidential elections in the U.S. demonstrating that social media have become increasingly important for politics (Wattal et al., 2010). Obviously, the instrumental use of social media in the political sphere can successfully adapt to the peculiar conditions prevailing during election periods, enabling politicians to come together and have discussions with their constituents and to deliver significant information about themselves. In particular, young people have begun to gain contact with political issues, using social media as communication platforms (Chen et al., 2009; Kushin and Kitchener, 2009).

Each social networking site has created different tools to connect users and to increase communication worldwide. Facebook and Twitter are two of the most popular social networking sites used by politicians to connect with voters. A strategy based on social media should take into account the different characteristics of each site in order to fully exploit the possibilities of communication. Facebook gives politicians during election campaigns the opportunity to create a dialogue with users and expand the reach of their campaign, and its use also offers the potential for targeted advertising. Twitter, in turn, allows candidates to continuously update the public on their activities with short messages, which offers another way for citizens to participate in discussions and communicate with politicians, as candidates can be connected with voters and bypass traditional media with their own posts that directly reach users. Each of these tools can improve the strategy regarding social media if they are harmonised and promote the message that each candidate promoted during his/her election campaign.

These online tools can prove to be very useful in an election campaign when used as channels through which continuous communication is achieved with the electorate, especially when it comes to the youngest segment of the population. One of the major advantages for politicians using social media is the opportunity afforded to them to present another image of themselves—one that is more personal and more accessible—in order to build stronger links with citizens. Social media therefore offer capabilities that traditional advertising does not include and a range of inexpensive advertising tools that contribute to interaction, participation and mobilisation of young people.

It is clear that social media offer the possibility of attracting potential voters, lowering the level of apathy on policy and increasing the level of political participation, but the degree of use by politicians and the degree to which people are interested in participating are not clear. With increasing political participation, a large part of the population interacts with their elected representatives, which could lead to a strong civil society, approaching in this way the ideal of active participation of the citizens and especially young people, who can play a role in political decisions (Gane and Beer, 2008).

The use of social media during election campaigns is now a key element to becoming or remaining competitive. The period of the elections also remains one of the most active periods and includes the most active users of social media in politics. These instruments of political marketing give politicians the opportunity to come into contact with a particular segment of the population when access to it may not be easy otherwise. This share includes the young and the growing number of digitally – skilled part of population (Anger and Kittl, 2011). Regarding Greece, the study of the online policy attracted the attention of researchers recently (Demertzis et al., 2005; Yannas and Lappas, 2005). The Internet was not part of the election campaigns of candidates in Greece until the local and municipal elections of 1998. At that time, only a very small number of candidates published texts online. Internet use in political campaigns was most prevalent in the 2000 parliamentary elections, and in 2004, the number of politicians using this new way of communicating doubled.

A study of the above literature and empirical research shows that there is a close connection between social media and political marketing, and to examine this association in order to understand how social media are used in modern politics, some research questions were developed. Specifically, this research focuses on the study of the use of Facebook and Twitter by politicians during an election period in order to attract voters. Additionally, it considers the differences arising from the use of Facebook and Twitter by politicians and also where the greatest interest lies in the content of their posts.

6.2.2 *Research questions*

In order to understand how social media are used in modern politics, three research questions were developed for this study, related to quantitative factors.

Q1: In which ways are Facebook and Twitter used so as to attract voters?
Q2: What differences arise in the use of Facebook and Twitter from the three politicians?
Q3: Is the use of both social media affected by the period seen in the context of approaching the day of the elections?

6.2.3 *Methodology*

This work aims to evaluate the use of social media in the political sphere as a means of marketing, disseminating information and encouraging political dialogue. For this purpose, we examine the way in which three politicians— Alexis Tsipras, leftist party; Fofi Gennimata, social democratic party; and Kyriakos Mitsotakis, conservative party—use their pages on Facebook and Twitter with the greatest efficiency and how they incorporate these new media as key tools for their political campaigns and tactics, causing the majority of the users' reactions. It should be noted that the choice of Kyriakos Mitsotakis was based on his presence on both Facebook and Twitter at that time, as Vangelis Meimarakis, who held the position of chairman of New Democracy, did not have accounts with both companies during the election period.

The research method that was followed was an analysis of user data in order to explore the way that social media consolidate the diffusion of the political messages and the substantial work that the politician does during the election period. The test period is the pre-election period in September 2015— namely from 30 August to 20 September. The special accounts examined are, for Facebook, Alexis Tsipras, Fofi Gennimata and Kyriakos Mitsotakis and, for Twitter, @AlexisTsipras, @FofiGennimata and @kmitsotakis.

6.2.4 *Social media selection*

The first step in collecting data was to determine the sources of data. Because of the soaring growth of use in recent years, the social networking sites and micro-blogging services with more use, Facebook and Twitter, provide an unprecedented amount of information associated with the political content. This contributed to their selection as the main sources of data from the area of social media.

6.2.5 *Period of data collection*

The recording of the data was held in November 2015. The data collection interval is defined as the campaign period from 30 August to 20 September 2015, a period during which social media played an important role in the communication campaign, allowing identification of differences in online behaviour policies. This period is divided into three parts:

* 30 August to 6 September 2015
* 7 to 13 September 2015
* 14 to 20 September 2015

6.2.6 *Research background*

The main objective of this research is to develop a political framework for the use of social media in political marketing, examining the users' reactions to the posts of politicians. In order to achieve the objectives of this study, a review of literature was carried out to identify the components of a framework for the implementation of social media strategies for political marketing.

6.2.7 *Coding*

The coding of data on posts to Facebook and Twitter is based in a large part on a survey conducted by Reuters Institute of Oxford University.

Politicians	Code list 1
Alexis Tsipras	1
Fofi Gennimata	2
Kyriakos Mitsotakis	3
Period	**Code list 2**
30 August to 6 September 2015	1
7 to 13 September 2015	2
14 to 20 September 2015	3

Genre (type of post)	Code list 3
Activities, meetings	1
Speeches (location and audience description)	2
Personal message (expression of feelings)	3
Political message (request for active participation in the elections)	4
Opinions (reference on political positions of the party)	5
Opinions (reference on political positions of opponents)	6
Interaction with followers/friends	7
Commenting on/analysing current affairs	8
Interviews in media	9
Other	98

Character	Code list 4
Positive	1
Negative	2
Neutral	3
Positive and negative	4

Thematics	Code list 5
Economy	1
Unemployment	2
Development	3
Education	4
Health	5
Administration	6
Social care	7
Immigration	8
Foreign affairs and defence	9

Form of post	Code list 6
Text	1
Photo	2
Video	3
Combination	4

6.2.8 *Data analysis*

After recording all posts of the three politicians from Facebook and from Twitter, as well as the users' reactions to them, the data were coded and initially transferred to Excel spreadsheet files. They were then moved into the SPSS statistics 20.0 programme for statistical processing and the production of graphs. The analysis included

1 A detailed presentation of the data via Facebook and Twitter, with the use of statistics. Frequencies and percentages appear for each of the variables.
2 A comparison of the use of Facebook and Twitter among the three politicians, which led to our finding the existence of a significant correlation between the independent variable of social medium—Facebook (1)

or Twitter (2)—and the dependent variables. Control of the significant correlation was done using chi-square tests. The hypothesis-testing procedure was used (with significance level was 0.05) for the following:

H0 = There is no correlation between the type of instrument and the dependent variables.
H1 = There is a correlation between the type of instrument and the dependent variables.

6.3 Findings

6.3.1 *Facebook*

Table 6.1 and Figure 6.1 present the frequency and the rate for each politician that participated in the survey. Fofi Gennimata has most of the usage, with 88%.

Table 6.2 and Figure 6.2 show the frequencies for the periods during which the three politicians made their posts on Facebook. Specifically, the period from 14 to 20 September 2015 had the largest share, 42.3%.

6.3.2 *Twitter*

Table 6.3 and Figure 6.3 present the frequency and the rate for each politician that participated in the survey. Alexis Tsipras has most of the usage, with 61.7%.

Table 6.4 and Figure 6.4 show the frequencies for the periods during which the three politicians made their posts on Twitter. Specifically, the periods from 7 to 13 September and from 14 to 20 September 2015 have the same frequency percentages, 36.1%.

6.3.3 *Comparison of the use of Facebook and Twitter*

The use of Facebook and Twitter among the three politicians was compared, and the existence of a significant correlation between the independent variable of social medium—Facebook (1) or Twitter (2)—and the dependent variables was found. Control of the significant correlation was done using

Table 6.1 Facebook usage by politician

	Frequency	Percentage	Cumulative Percentage
Alexis Tsipras	12	8.5	8.5
Fofi Gennimata	125	88.0	96.5
Kyriakos Mitsotakis	5	3.5	100.0
Total	142	100.0	

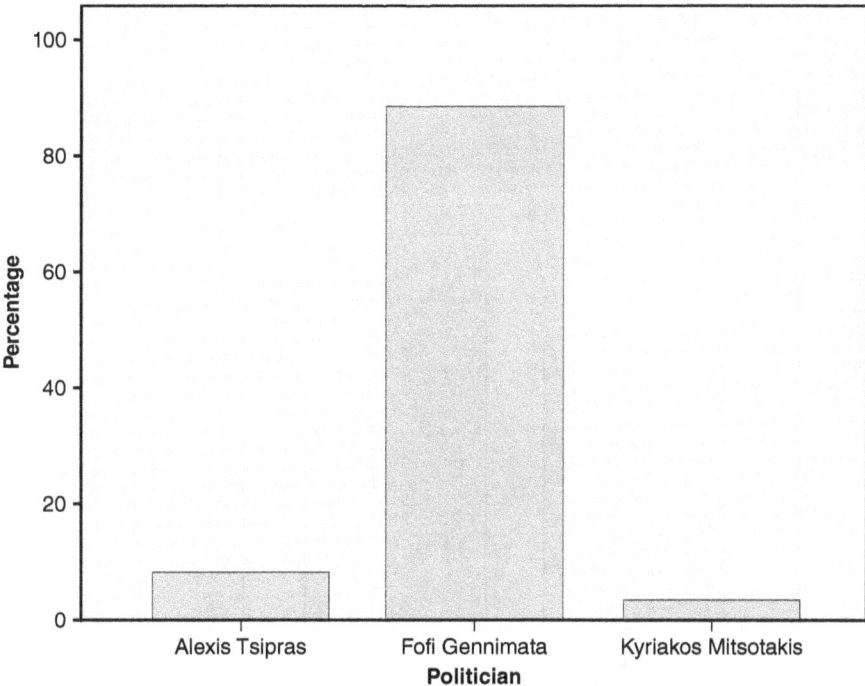

Figure 6.1 Facebook usage by politician

chi-square tests. The hypothesis-testing procedure was used (with a significance level of 0.05) for the following:

H0 = There is no correlation between the type of instrument and the
      dependent variables.
H1 = There is a correlation between the type of instrument and the
      dependent variables.

Facebook shows a higher rate of the positive posts compared to Twitter—
i.e., 71.8% and 52%, respectively—while Twitter shows more than double the
negative posts compared to Facebook—i.e., 25.2% and 10.6%, respectively.
(See Table 6.5 and Figure 6.5.)

Table 6.2 Period

	Frequency	Percentage	Cumulative Percentage
30 August to 6 September	35	24.7	24.7
7 to13 September	47	33.1	57.7
14 to 20 September.	60	42.3	100.0
Total	142	100.0	

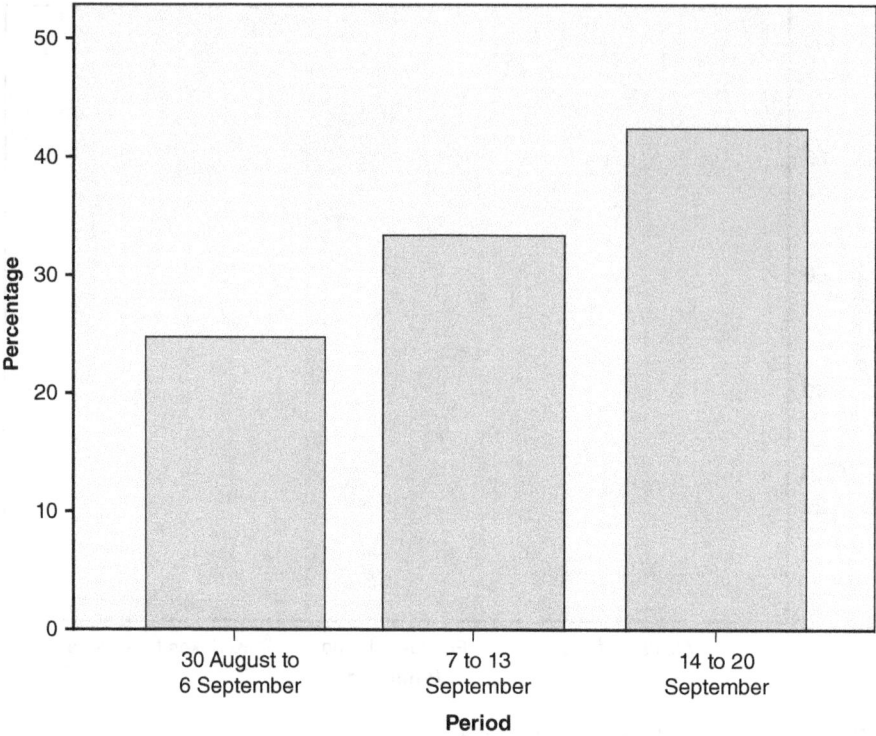

Figure 6.2 Period

On Facebook, photos and videos are the main types of posts used, with rates of 45.8% and 28.9%, respectively, and the main type used on Twitter is simple texts, with a rate of 85%. (See Table 6.6 and Figure 6.6.)

From Table 6.7 and Figure 6.7, the conclusion is drawn that Twitter posts were related to the greatest extent to political messages and opinions on partisan opponents, with rates of 24% and 23.4%, respectively. By contrast, the smallest percentage, 2%, involved current affairs analysis and comments. On Facebook, posts were related to the greatest extent to speeches and political messages, with rates of 18.3% and 15.5%, respectively. By contrast, the smallest percentage, 2.1%, involved current affairs analysis and comments.

Table 6.3 Twitter usage by politician

	Frequency	Percentage	Cumulative Percentage
Alexis Tsipras	499	61.7	61.7
Fofi Gennimata	203	25.1	86.8
Kyriakos Mitsotakis	107	13.2	100.0
Total	809	100.0	

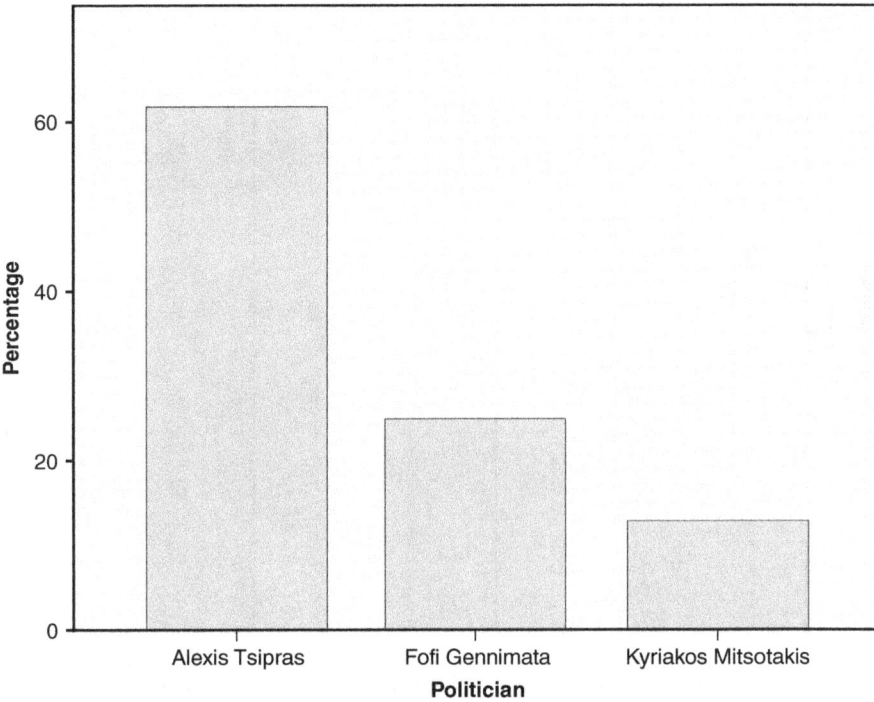

Figure 6.3 Twitter usage by politician

The results in Table 6.8 and Figure 6.8 show that economic issues dominate the content of postings in both media, and especially on Twitter, with rates of 47.9% and 26.1% on Twitter and Facebook, respectively. Issues of foreign policy and defence received a very small percentage of posts, 0.7%, on Twitter, while on Facebook no reference to this issue is made.

6.3.4 *Results discussion*

This study presented a comprehensive analysis of the use of Facebook and Twitter by three politicians, Alexis Tsipras, Fofi Gennimata and Kyriakos Mitsotakis. The findings are very interesting, especially in disputes arising from how to use Twitter compared with Facebook, as the results show that

Table 6.4 Period

	Frequency	Percentage	Cumulative Percentage
30 August to 6 September	225	27.8	27.8
7 to 13 September	292	36.1	63.9
14 to 20 September	292	36.1	100.0
Total	809	100.0	

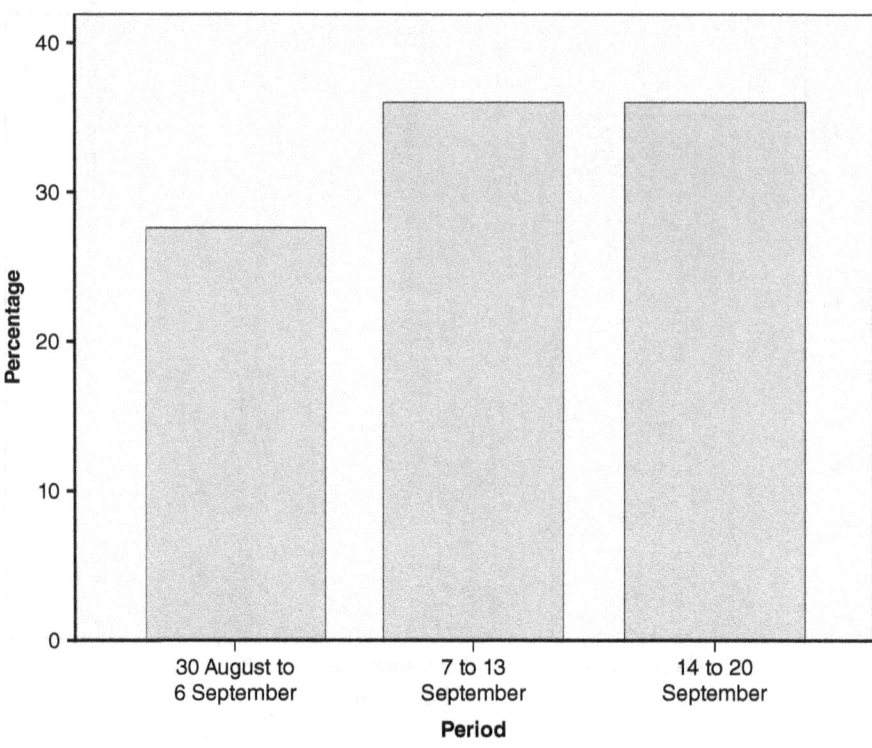

Figure 6.4 Period

Table 6.5 Character

			Positive	Negative	Neutral	Positive and Negative	Total
					Character		
Social Media	**Facebook**	Count	102	15	13	12	142
		%	71.8%	10.6%	9.2%	8.5%	100.0%
	Twitter	Count	421	204	74	110	809
		%	52.0%	25.2%	9.1%	13.6%	100.0%
Total		Count	523	219	87	122	951
		%	55.0%	23.0%	9.1%	12.8%	100.0%

Chi-Square Tests

Value	df	Asymp. Sig. (2-sided)
22.360[a]	3	0.000
24.409	3	0.000
9.800	1	0.002
951		

[a] 0 cells (0.0%) have an expected count less than 5. The minimum expected count is 12.99.

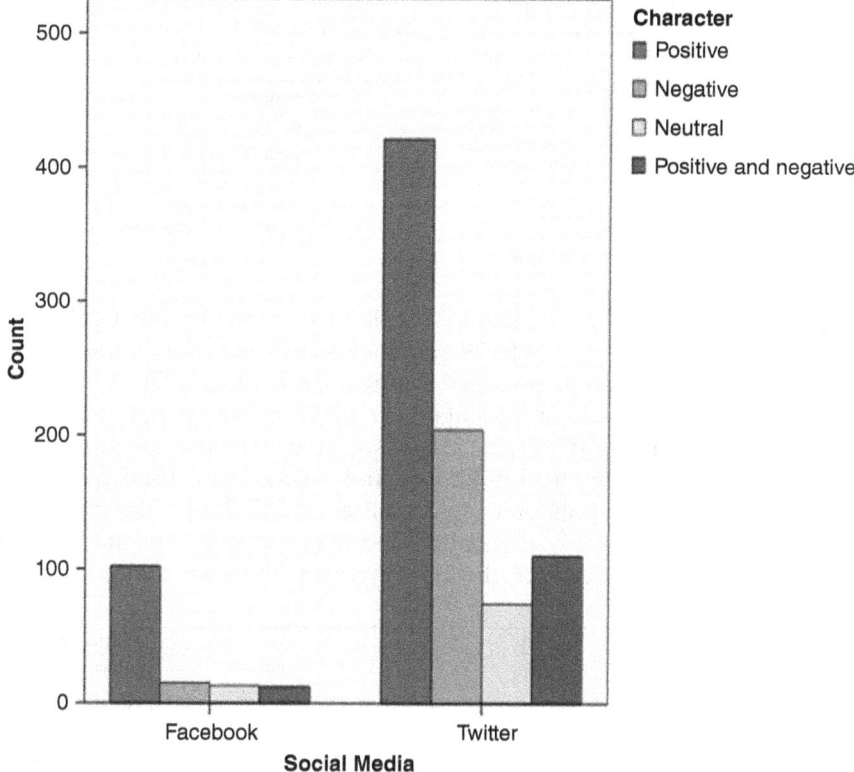

Figure 6.5 Character

the three politicians did not similarly use these two social media. As for Facebook, this is related to the degree of use. As for the general number of tweets by each politician, we observe that Alexis Tsipras has the largest number of tweets in relation to the other two candidates, followed by Kyriakos Mitsotakis, while on Facebook Fofi Gennimata is the most frequent user. As to the time periods, the posts in both social media increase between the

Table 6.6 Post type

			Photo	Video	Text	Combination	Total
Social Media	**Facebook**	Count	65	41	18	18	142
		%	45.8%	28.9%	12.7%	12.7%	100.0%
	Twitter	Count	96	2	688	23	809
		%	11.9%	0.2%	85.0%	2.8%	100.0%
Total		Count	161	43	706	41	951
		%	16.9%	4.5%	74.2%	4.3%	100.0%

Chi-Square Tests

Value	df	Asymp. Sig. (2-sided)
413.267ª	3	0.000
344.513	3	0.000
139.377	1	0.000
951		

ª 0 cells (0.0%) have an expected count less than 5. The minimum expected count is 6.12.

first and third periods. It is worth noting that use of Twitter during the periods 7 to 13 September and 14 to 20 September 2015 was exactly the same.

Additionally, Facebook presents a higher percentage of positive posts compared to Twitter, which presents more than double the number of negative posts. Regarding the form, it is associated with the social networking medium used. Specifically, photos and videos were used mainly on Facebook by the three politicians to promote their positions, the supporters of the campaigns or family, depending on the message they wanted to send to their voters, and on Twitter, the main type was the simple text.

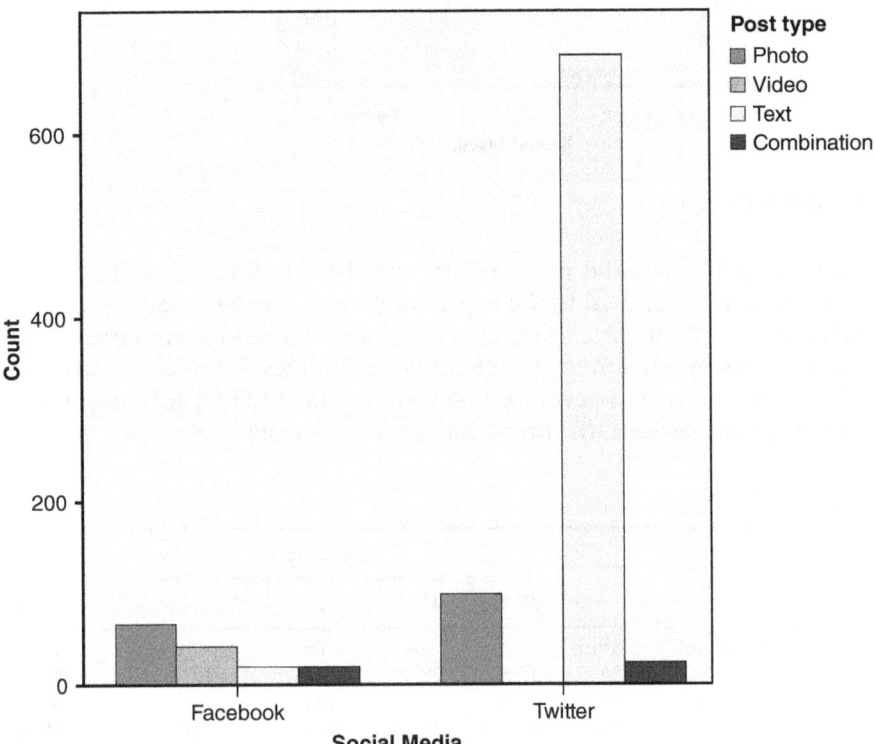

Figure 6.6 Post type

Table 6.7 Genre of post

Genre of post		Social_Media		
		Facebook	Twitter	Total
Activities, meetings	Count	18	19	37
	%	12.7%	2.3%	3.9%
Speeches (location and audience description)	Count	26	35	61
	%	18.3%	4.3%	6.4%
Personal message (expression of feelings)	Count	16	162	178
	%	11.3%	20.0%	18.7%
Political message (request for active participation in the elections)	Count	22	194	216
	%	15.5%	24.0%	22.7%
Opinions (reference on political positions of the party)	Count	15	147	162
	%	10.6%	18.2%	17.0%
Opinions (reference on political positions of opponents)	Count	10	189	199
	%	7.0%	23.4%	20.9%
Interaction with followers/friends	Count	14	15	29
	%	9.9%	1.9%	3.0%
Commenting on/analysing current affairs	Count	3	16	19
	%	2.1%	2.0%	2.0%
Interviews in media	Count	18	32	50
	%	12.7%	4.0%	5.3%
Total	Count	142	809	951
	%	100.0%	100.0%	100.0%

Chi-Square Tests

Value	df	Asymp. Sig. (2-sided)
141.067[a]	8	0.000
116.101	8	0.000
1.783	1	0.182
951		

[a] 2 cells (11.1%) have an expected count less than 5. The minimum expected count is 2.84.

As to the type of post, it is influenced by the type of social networking medium, with the bulk of posts in both media focusing on promotion of their speeches: mainly to promote positions against party rivals on Twitter and to promote political messages on Facebook. Also the posts of the three politicians expressing their own independent opinions on various topics are categorised according to their thematic content, and the results show that economic issues dominated in both social media.

6.4 Theory recommendations

Previous studies have shown that in recent years social media have become an important channel of marketing in the political sphere and allow political players and voters to have direct interactions. Therefore, the political

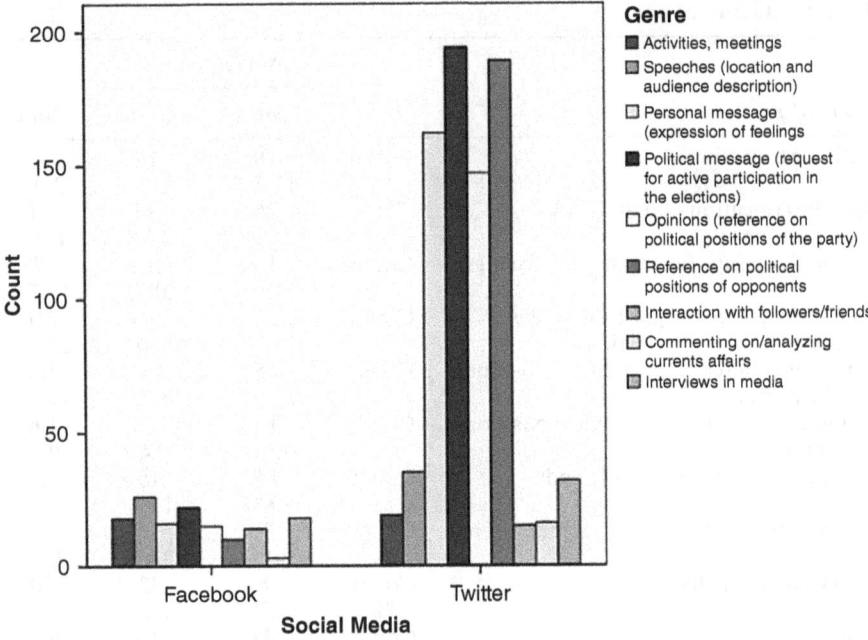

Figure 6.7 Genre of post

Table 6.8 Themes

Cross-Tabulation of Themes and Social Media		Social_Media		
		Facebook	Twitter	Total
Economy	Count	6	190	196
	%	26.1%	47.9%	46.7%
Unemployment	Count	1	10	11
	%	4.3%	2.5%	2.6%
Development	Count	4	49	53
	%	17.4%	12.3%	12.6%
Education	Count	2	18	20
	%	8.7%	4.5%	4.8%
Health	Count	3	8	11
	%	13.0%	2.0%	2.6%
Administration	Count	3	37	40
	%	13.0%	9.3%	9.5%
Social care	Count	2	50	52
	%	8.7%	12.6%	12.4%
Immigration	Count	2	29	31
	%	8.7%	7.3%	7.4%
Foreign affairs and defence	Count	0	6	6
	%	0.0%	1.5%	1.4%
Total	Count	23	397	420
	%	100.0%	100.0%	100.0%

Chi-Square Tests

	Value	df	Asymp. Sig. (2-sided)
Pearson Chi-Square	14.800[a]	8	0.063
Likelihood Ratio	10.630	8	0.224
Linear-by-Linear Association	1.151	1	0.283
N of Valid Cases	420		

[a] 8 cells (44.4%) have an expected count less than 5. The minimum expected count is 0.33.

activities could be more transparent, and citizens could engage more actively in political decision-making processes. However, until now, the possibilities for political discussions in social media have been at an early stage, with politicians struggling to adapt to these new conditions.

Many of the protagonists of the political scene try to appropriate this new logic of communication through new networks, while some remain tied to traditional methods of communication with voters, especially in election periods. This new form of marketing, of course, is quite different and requires other skills from both politicians and voters. Politicians should be familiar with the technology, should know how to effectively use these social networking platforms and should also be familiar with technological innovations. At this point, we should emphasise that when politicians use these social media

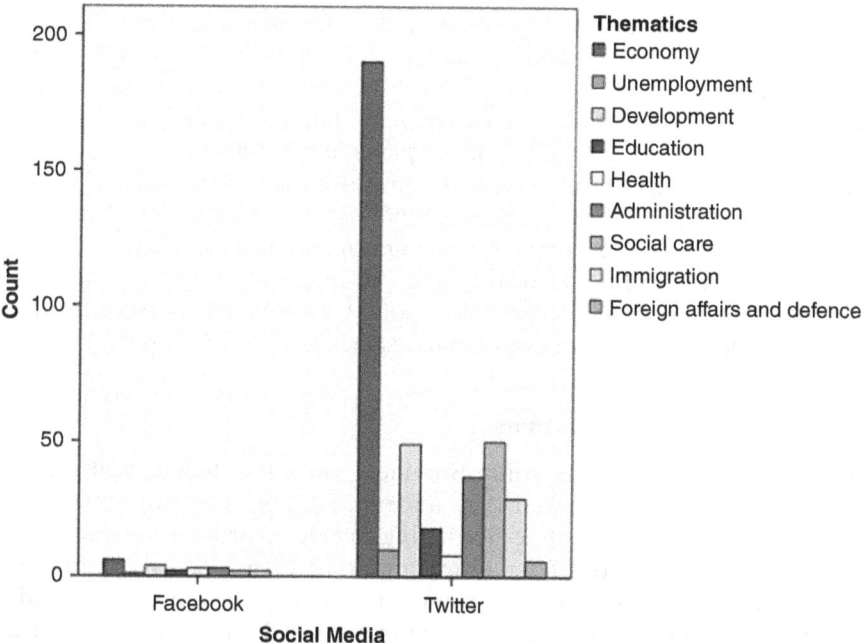

Figure 6.8 Themes

networks, essentially they send the information through a channel that is not filtered or controlled by traditional media. In this way, they are able to carry their own political agenda, not content to be censored or filtered.

For citizens, increased political participation through the efforts of social media can manifest itself in the form of increased fundraising, greater participation by volunteers and also increased knowledge about the candidates (Partridge, 2011). Political parties and politicians, and especially the candidates, that are able to effectively use social media can reach voters faster and cheaper—and, some argue, more effectively—than those who do not make much use of both these online tools. As the use of social media has increased in the United States, public opinion is now influenced by the candidates, their campaigns, their opinion leaders and their posts on social networking sites, which are then shared by the followers of social networks, ensuring in this way that promoted ideas are spread.

Thus, communication and marketing through social media remain the primary reason for their use in politics today, as they serve as an additional tool for marketing and information transmission. However, this trend is evolving to become more complex, and social media have begun to be used as an interaction and mobilisation tool on the modern political chessboard. The important thing today is for a politician to approach social media as promotional tools of political positions. The social scale in which social media is now addressed is very big; thus, in case of error, the deterioration of a politician's image could be immense. A new era has dawned, in which political time has shrunk considerably and within minutes, both message and sender are evaluated. The upcoming elections in the U.S. will be an electoral battle in which social media will have more influence than ever; compared to 2012, the number of Twitter users has almost doubled, and the number of active Instagram users has tripled (Rossi, 2015).

Social media are constantly evolving, thus resulting in new uses and applications. The most important for the future of this new field of political marketing is the use of social media by governments and political actors to continue to move from the traditional transmission of information to the emerging trend of exchange and interactive cooperation for the benefit of the information society, development policy and public participation and supervision.

6.5 Practical recommendations

It is worth noting that this study provides political actors a useful and practical diagnostic tool; when they understand how they can better use it, they can 'build' their online marketing profile in order to attract voters. So far the possibilities for political discussions in social media are at an early stage, and politicians are struggling to adapt to these new conditions. Social media are constantly evolving, thus resulting in new uses and applications. As part of this development, there is an important distinction between the simple use and the active use of social media. The opportunities for information and cooperation provided by these political instruments are vital to understanding how and why these marketing tools can and should be used in politics.

Additionally, social media recreate dialogues and online discussions that can radically change the form of political marketing. If these marketing tools are used effectively, they have the power to remove political messages from the old model of media and to promote a more direct form of communication. These social media can act as a 'mood barometer' and also as a useful tool for exploring new areas where politicians can gain access to younger audiences. The ability to call on the public to participate in dialogues on policy issues, the fact that the media themselves are part of current discussions on the Internet in a targeted manner and the fact that politicians might coordinate such discussions may give them a chance to prevail in the middle of tense political situations. In short, politicians need to become masters of social media—not only to know how to better collect more reliable information to comply with the decision-making deadlines but also to know how to exercise the greatest possible influence on public dialogue through these tools. The political landscape in the digital age becomes more complicated, and in a world of 'likes' and 'hashtags', the political protagonists cannot avoid adapting to changes that social media offer.

6.6 Limitations and recommendations for future research

This study is subject to some restrictions, which the researcher must identify and which can be used to lead to further research. The period during which the use of Facebook and Twitter by politicians was considered was too short, as the campaign period lasted about three weeks. We also saw a vision of the use of social networks, but this study can be extended to develop a survey, since social media have enabled citizens to have access to political websites at an unprecedented level. The focus of this research can be shifted to the study of the users' reactions towards the politicians' posts. Specifically, the users' reactions could be captured on the basis of coding posts by themes, such as health, economy, education, foreign policy and immigration, in order to identify the issues that are high on the political agenda and cause the greatest interest of users.

Additionally, this research can be extended to focus on the use of a new service that was introduced on Facebook. The addition a few months ago of application now allow users to express their feelings through emoticons on posts and represent 'love, laugh, smile, shock and anger'.

References

Anger, I., and Kittl, C. (2011). Social media in European governmental communication. In *Proceedings of the 11th European Conference on E-government* (pp. 43-52). Reading, UK: Academic Publishing.

Baines, P., Harris, P., and Lewis, B. (2002). The political marketing planning process: Improving image and message in strategic target areas. *Marketing Intelligence and Planning, 20*(1), 6-14. DOI:10.1108/0263450021041410

Bakker, T. P., and de Vreese, C. H. (2011). Good news for the future? Young people, Internet use, and political participation. *Communication Research, 38*(4), 451-470.

Baumgartner, J. C., and Morris, J. S. (2010). MyFaceTube politics: Social networking web sites and political engagement of young adults. *Social Science Computer Review, 28*(1), 24-44. DOI:10.1177/0894439309334325

Bryer, T. A., Callen, J. C., Eikenberry, A. M., Garrett, T. A., Love, J. M., Miller, C. R., Stitch, B., and Wickstrom, C. (2010). Public administration theory in the Obama era. *Administrative Theory and Praxis, 32*(1), 118–122. DOI:10.2753/ATP1084-1806320107

Chen, D., Tang, J., Li, J., and Zhou, L. (2009). Discovering the staring people from social networks. In J. Quemada (Ed.), *Proceedings of the 18th International Conference on World Wide Web* (pp. 1219-1220). New York: Association for Computing Machinery.

Dean, D., and Croft, R. (2001). Friends and relations: Long-term approaches to political campaigning. *European Journal of Marketing, 35*(11/12), 1197-1216. DOI:10.1108/EUM0000000006482

Demertzis, N., Diamantaki, K., Gazi, A., and Sartzetakis, N. (2005). Greek political marketing online: An analysis of Parliament members' web sites. *Journal of Political Marketing, 4*(1), 51-74. DOI:10.1300/J199v04n01_04

Dermody, J., and Scullion, R. (2001). An exploration of the advertising ambitions and strategies of the 2001 British general election. *Journal of Marketing Management, 17*(9-10), 969-987.

Gane, N., and Beer, D. (2008). *New media.* Oxford: Berg Publishers.

Gilmore, J. (2012). Ditching the pack: Digital media in the 2010 Brazilian congressional campaigns. *New Media and Society, 14*(4), 617-633. DOI:10.1177/1461444811422429

Gulati, G. J., and Williams, C. B. (2010). Congressional candidates' use of YouTube in 2008: Its frequency and rationale. *Journal of Information Technology and Politics, 7*(2/3), 93-109. DOI:10.1080/19331681003748958

Harris, P., Lock, A., and O'Shaughnessy, N. (1999). Measuring the effect of political advertising and the case of the 1995 Irish Divorce Referendum. *Marketing Intelligence and Planning, 17*(6), 272-279. DOI:10.1108/02634509910293089

Hong S, Nadler D (2011) *Does the Early Bird Move the Polls? The use of the social media tool 'Twitter' by U.S. politicians and its impact on public opinion.* In: Proceedings of the International Conference on Digital Government Research

Hughes, A., and Dann, S. (2009). Political marketing and stakeholder engagement. *Marketing Theory, 9*(2), 243-256.

Kim, W., Jeong, O.-R., and Lee, S.-W. (2010). On social web sites. *Information Systems, 35*(2), 215-236.

Kushin, M., and Kitchener, K. (2009). Getting political on social network sites: Exploring online political discourse on Facebook. *First Monday, 14*(11). DOI:10.5210/fm.v14i11.2645

Lock, A. and Harris P. (1996), Political Marketing, *European Journal of Marketing, 30*(10), 14-24

Murray, K. E., and Weller, R. (2007). Social networking goes abroad. *International Educator, 16*(3), 56-59.

Ormond R. (2012). *Defining Political Marketing,* Management Working Papers.

Partridge, K. (2011). *Social networking*. New York: H. W. Wilson.

Peng, N., and Hackley, C. (2007). Political marketing communications in the UK and Taiwan. *Marketing Intelligence and Planning, 25*(25), 483-498. DOI:10.1108/02634500710774950

Powell, L., Richmond, V. P., and Williams, G. C. (2011). Social networking and political campaigns: Perceptions of candidates as interpersonal constructs. *North American Journal of Psychology, 13*(2), 331-342.

Rossi, B. (2015). *Will the 2016 presidential election be decided on social media?* Information Age, http://www.information-age.com/it-management/strategy-and-innovation/123459883/will-2016-presidential-election-be-decided-social-media. Accessed 6 December 2015.

Shama, A. (1976). Marketing the Political Candidate, *Journal of the Academy of Marketing Science, 4*, (4), 764-777.

Towner, T. L., and Dulio, D. A. (2011). An experiment of campaign effects during the YouTube election. *New Media and Society, 13*(4), 626-644. DOI:10.1177/1461444810377917

Wattal, S., Schuff, D., Mandviwalla, M., and Williams, C. (2010). Web 2.0 and politics: The 2008 U.S. presidential election and an e-politics research agenda. *MIS Q, 34*(4), 669-688. http://www.jstor.org.ezproxy3.lib.le.ac.uk/stable/25750700?seq=1#page_scan_tab_contents.

Yannas, P., and Lappas, G. (2005). Web campaign in the 2002 Greek municipal elections. *Journal of Political Marketing, 4*(1), 33-50. DOI:10.1300/J199v04n01_03

Zhang, W., Johnson, T. J., Seltzer, T., and Bichard, S. (2010). The revolution will be networked: The influence of social networks on political attitudes and behaviors. *Social Science Computer Review, 28*(1), 75-92. DOI:10.1177/0894439309335162

7 The brain economy

*Tatjana Dragičević Radičević
and Milica Nestorović*

7.1 Introduction: Global view of creating the new forms of economy

Determined by the external environment and with its complexity of economic relations, contemporary society presents a challenge for scientists and researchers of different scientific fields. As a result, they are facing a wide scope of new dilemmas, especially those brought about by globalisation, a phenomenon that has assumed great importance in the 20th and 21st centuries. There are different views of globalisation in terms of whether it is a product of a modern society or a dialectical process that has changed the instruments and resources of implementation. The authors of this chapter support the attitude that globalisation is the realistic and evolutionary process that represents the unconditional flow of radical changes in society, technology, economics, politics, religion, culture and media, as well as in many other areas. Therefore, this chapter synthetises the results of many years of research in the fields of socio-economic relations development and globalisation.

Many indicators, including qualitative, quantitative, technological, economic and other determinants, show that contemporary changes are significantly different—that is, more complex—than they have been before. In this regard, Peter Drucker (1994) believes that they are even more radical than the changes that indicated the beginning of the Second Industrial Revolution in the mid-19th century or the structural changes that caused the Great Depression of 1929-1933 and the Second World War. Today, mankind is believed to have the technologies in the fields of communications and transport essential for the formation of a global world economy (Dragičević Radičević, Cvetković Bogavac and Gavrilović, 2012).

The internationalisation of capital, accompanied by the rapid development of technology, has enabled intensive global integration of all elements of economic life. It is exactly the correlation, causality and dependence of the market economy and technological progress that make one of the most important and strongest driving levers of modern civilisation. It is the ruthless market competition in which the winner is the one with a highest profit and best technology, which indicates the strong technological determinism of globalisation.

Studying the phenomenon of civilisation and the evolutionary phases of the economy, we could say that globalisation and economic interests of individuals and the states are directly conditioned. In the 18th century, in his book *The Wealth of Nations*, Adam Smith underlined the importance of the individual and the common good. In the conclusion of the globalisation aspects analysis, there seems to dominate the fact that there is a synergistic effect of various aspects of globalisation (economic, social, technological, cultural, etc.) with the economic globalisation, based on fast-evolving technology, representing primus inter pares among all globalisation aspects.

It is often pointed out that globalisation is a dialectical process in social relations; that is, the process involves the interaction of opposite sides in the society. From an entirely realistic point of view, the process of globalisation with its accompanying aspects brings both advantages and disadvantages for all countries, irrespective of each particular country's level of development. The very process of globalisation, which has changed its forms and manifestations over the centuries, is not problematic in itself. On the contrary, the problem lies in those who run it (Nestorović, 2016).

The contradictions that globalisation brings make the modern infrastructure of the new world society. The process of globalisation has evolved through the following four periods (Dragičević Radičević, 2007, 42):

I   Protoglobal orders
II   Renaissance
III   Colonial conquests
IV   Mega globalisation

Throughout the evolutionary process, the instruments in global trends have been changing in accordance with the development of social consciousness and social being, while the goal has evolved and become increasingly materialised in the form of profit. More precisely, the well-being brought about by globalisation in the form of global connectivity, the transcendence of national boundaries and the free flow of ideas, information and knowledge has created a megalomaniac power, which, in turn, is directly connected with profit. The primary evolutionary development was accomplished through the civilisational, religious and spiritual connections, while the latest evolutionary development, that of mega globalisation, represents the reflection of the contemporary social, political and economic trends (Dragičević Radičević, 2016).

If we should give the paradigm of the process of globalisation, we could say that the integration of capital markets, technology, information, communication and knowledge represents the foundation of this process. This whole complex process results in the financial strength of transnational and multinational corporations in developed countries and in the liberalisation of markets. However, such effects and market liberalisation benefited only the developed markets, which resulted in the undeveloped and developing countries necessarily initiating the transition processes, moving from the current to a new, desirable condition with the aim of preserving economic stability. These processes differed from country to country.

Modern globalisation (Jakšić, 2003) is taking place predominantly on the wings of technology. Its expansion and acceleration are unstoppable and result from the created multiple sources of knowledge turned into various technologies that effectively control not only the economic but also all other trends. The Industrial Revolution and technological advances have reduced production costs, introduced new production capacities and enabled the creation of new products. In a global economy, the factors of production, goods, services, natural resources, capital, technology, labour and information are highly mobile. Manufacturers transfer their production plants to the cheapest places—the less-developed and developing countries—while speculators make profit by shifting these factors from the cheaper places to those where they are more expensive. The national market is becoming less so; it turns into a part of a single global market instead.

Globalisation is characterised by the interconnectedness of countries into a multidimensional network of economic, social, cultural, political and other ties. Globalisation is not related only to the economic sphere but also to the areas of politics, technology, science, religion, culture, entertainment, media and so on. It has flooded all areas of individual and collective existence. As a comprehensive process, globalisation combines different perceptions, aspects and views on what is happening in the world. Thus, this process connects economists, political scientists, culturologists, technologists, ideologists, ecologists, and other participants in the world market. In that sense, it is not surprising that some authors observe it through the prism of modernity—one of them being Anthony Giddens, the world's leading sociologist, who believes that globalisation is in many aspects not only a new but also a revolutionary process, considered to be as much a political, technological and cultural process as it is an economic process (Fischer, 2003).

Radical changes in generally accepted standards and rules of life, ranging from the social, environmental, economic, legal, political and cultural to the international, are the results of the use of sophisticated technology (IT, telecommunications, industrial, energy, medical, genetic innovation and other high technology). Basically, the privilege of the ownership and usage of these forms of innovation is reserved for strong and developed economies and is unacceptable for the poor, small and undeveloped countries.

The other side of this ambivalent process leads to the appearance of unemployment, redundancy and more frequent fluctuations in population from undeveloped to developed regions. These are the result of automatisation and a higher level of production efficiency due to the 'replacement' of more workers with one machine. Manual work has given way to automatisation, but in the sphere of knowledge dissemination, intellectual work has become an 'expensive commodity'. There is an increasing need for people to participate in continuous education, open new horizons and develop new forms of gathering and acting, which gradually create a new system of societies, bringing about the changes in the form of economy.

Investment in knowledge results in the conquest of new markets and innovations in a variety of fields, such as new and cheaper raw materials. However, most innovations are in the domain of technology. Accumulating existing knowledge and information and gaining new knowledge and information (not easily available to everyone) create the increasing interdependence between national economies and the world economy.

As already stated, the information revolution, as a part of long-term globalisation flows, has led to radical changes in society, technology, economics, politics, religion, culture, media and many other areas. From the economic and social points of view, globalisation has initiated the transition processes in all the above-mentioned areas, as values now acquire a new dimension. New global norms and values are being established that overcome national boundaries. As one of them, the change of established standards and rules related to information and knowledge stands out; it is the change from individual to collective forms of their definition and usage. In a parallel process, new forms of economy are created by the formation of innovative values in different spheres of the lives of individuals and the society.

The assumption is that great social changes, caused by dialectical developments and the globalisation in society, lead to complex changes of socio-economic forms as well, which essentially observe the individual and/or society as a whole. The dialectical process in the form of the Hegelian 'negation of the negation' indicates that all negative effects of today's social and economic forms lead to the creation of a new and higher form of economy.

7.2 Methodology research design

The methodological framework of the chapter is based on the authors' long-term research work (from 2000 to 2016) in the field of socio-economic relations development and globalisation. The research was based on the deductive principle of conclusion, qualitative theoretical analysis with the elements of description and the correlation of different economic and social variables, as well as Internet network interactive method. Data collection and analysis was carried out on the basis of relevant literature (research of reviewed scientific papers, books, electronic scientific research databases and other scientific editions), which studies the period of the scientific approach to the development of economic thought (18th-21st centuries), as well as the determinants of social changes by accelerating the effects of globalisation on social changes in the 20th and 21st centuries. Each methodological phase in this chapter is the result of the synthesis of individual research results, obtained at different time intervals during the research period.

In the first part of the chapter, the methodological framework was set up to define the dependent variables as instruments in the establishment of different forms of socio-economic relations by correlating relations between globalisation and socio-economic trends. Such an assumption arises from the defined hypothesis of research: The assumption is that the great social changes caused by dialectical developments and globalisation in society also lead to complex changes of the socio-economic forms, which essentially observe the individual and/or society as a whole. The dialectical process in the form of the Hegelian 'negation of the negation' indicates that all the negative effects that today's socio-economic forms have contribute to the formation of a new and higher form of economy.

The second segment of the research is based on a deductive-analytical method, which results in defined forms of socio-economic development. In the analysis and comparative qualitative method, as a result of the symbiosis and dialectical process of the previous phases, in this phase of the research a new form of economy was found, the so-called brain economy.

The third part of the research methodically corresponds to the analysis of determinants in socio-economic models, carried out by comparing the economic importance of information and knowledge in correlation with the social variables of individualism and collectivism.

The last segment of the research focuses on concluding observations that confirmed the hypothesis and contributed to the strengthening of scientific thought in the direction of correlational relations of globalisation and forms of economic development.

7.3 Economic and social development

The goals of an economy involve finding methods, mechanisms, instruments and goals in the search for the optimum in the rational use of resources, on the one hand, and maximum satisfaction of the needs of an individual and society, on the other. Finding the reconciliation between these two extremes, in a sense, represents the Gordian knot, whose solution has been searched for ever since the ancient philosophers.

The polarity of the motives and consequences accelerated in modern society when was globalisation reflected primarily in the economic dimension. However, the common denominator of all social and economic stages of development is the concept of individualism vs. collectivism. That conflict produces an evolutionary 'ideology', characterised by the growing social stratification and the reification of the individual. The socio-economic environment of the individual is quantified, with a complete ignorance of the qualitative role of the individual. In the 18th century, in 'moral arithmetic', Bentham underlined the complexity of this phenomenon, with the motives and consequences defined in terms of the reflections on the society and the individual. The determinant of success was measured by collective, national interests, with the consequences primarily reflecting on individuals. The Machiavellian principle became the global principle of *Finis santificato media*.

Today, we can witness that the Hegelian dialectic 'negation of the negation' has been confirmed because we are at a new crossroads again, brought upon us by the accelerated effects of social and economic determinants and triggered by the contemporary framework of globalisation but also by the social phenomena before that, such as the Industrial Revolution and the information technology revolution. Thus, the future is reflected in the new dialectical and the higher level of socio-economic development. In economics, this idea of a holistic approach is perhaps most closely supported by Manfred Max Neef, who says that 'the economy must serve people, and not the other way round' (1991: 32-33).

In the context of the previous understanding of the socio-economic evolution and from the perspective of the individualism vs. collectivism aspect, four socio-economic forms could be defined:

  I   Economy of an individual
  II Economy of scale
  III Economy of information and knowledge
  IV Brain economy

Each of them is the result of the social laws under which it was established and basic economic principles (production, distribution, exchange and consumption). Economy of an individual is the simplest form, with a minimum role of materialisation. However, it is embodied in all other economies and has led to the necessity of creating the brain economy by developing the dialectic 'negation of the negation'.

7.3.1 *I Economy of an individual*

In his book *Escape from Freedom*, Erich Fromm emphasises the key determinant in the analysis of socio-economic forms: 'Social history of man began with his rise from the state of unity with the nature to the consciousness of oneself as an entity which has separated from the surrounding nature and people' (1983: 41). In this definition, Fromm underlines the existence of the economy of an individual. An expression of this thesis can also be found in Friedrich Engels's (1945) *Anti-Duhring*, where he points out that in the tribal or rural municipality with the common ownership of land, there was equal distribution, and in this regard, he emphasises distribution as a key factor in the stratification of society and the separation from 'the state of unity', as claimed by Fromm later. Production and exchange are not the factors accelerating individualistic disintegration; it is the consequence of distribution.

Production and exchange limited to the local boundaries and a low degree of specialisation and technological development are the main characteristics of this form of socio-economic development. It was based on the exploitation of available natural resources and could be therefore called 'man (the individual) playing with nature'. At this stage of the economy, the predominantly used factors were those of labour and land, so class stratification was minimal and distribution was carried out to the satisfaction of each individual. The introduction of coins as a means of exchange brought about growing changes in distribution and social stratification, so that the natural laws that favour equal satisfaction of all individuals were spontaneously replaced.

'The clan-based society maintained the relative equality as long as there was a low level of production, i.e. as long as the communities spent their products' (Bošnjak, 1960: 15). After the dissolution of the tribal society, the slave society was created, which meant the increasing development of production and exchange and hence the stratification based on distribution, which was primarily reflected in property relations and not in blood relations, as was the case before. On the one hand, a land market was formed, and on the other, a cheap labour market arose in the form of slaves. However, although materialisation and quantification increasingly gained importance, the human being (an individual), as a knowledge being, was still the subject of studies, primarily among the Greek philosophers.

In the era of feudalism, materialisation and quantification became even more important, although the primacy was still on the land and the use of labour. Some philosophers and economists consider feudalism to have been the dawn of the capitalist era.

The Renaissance brings new directions in the development of socio-economic thought. This era laid the foundations of new progress: natural, scientific and technical. A new civil society was developed, which considered man to be the individual with his own ideas and work. Among the scientists, Giordano Bruno was the first to underline the desire of man to rise from the unity with nature to rule over nature (Filipović, 1956). Tommaso Campanella points to a 'social contrast between a minority who worked and the majority who lived only for the enjoyment' (Filipović, 1956: 87).

In his work *Utopia*, Thomas More writes about prosperity and poverty and discusses the issue of the common good, which was not fair in the distribution system of the time.

> I'm so, while observing thoughts and think about countries that exist in the world today, all I see, by God, is just a conspiracy of the rich, who are supposedly on behalf of and in the name of national interests, fight only for their personal interests. Devising and inventing all possible means, how to first and without risk to preserve all good, which has been fraudulently acquired, and then, how will with the lowest price rent and take advantage of the work and effort from the poor. As soon as the rich decide that gimmick should be implemented by the interest of the community, and therefore on behalf of the poor, they immediately become laws. (1952: 100)

Previous arguments suggest that in this period the development of socio-economic relations resulted in a growing social stratification and less fair distribution, the alienation of work results from their creators and the appropriation of surplus value by the owners of capital, primarily land. However, the dialectical 'negation of the negation' led to a higher level of socio-economic development and man's consciousness of this development, as well as the possibility to subdue the nature instead of being subdued by it.

7.3.2 *II Economy of scale*

Between the 16th and 18th centuries, the capitalist society was gradually built. The invention of the steam engine in 1764 initiated a sudden development of the society in the field of engineering and technology, which accelerated the socio-economic structure. This also increased the creation of surplus value, which is the main generator of unfair distribution. Social benefit and economic goals became a starting hypothesis, first proclaimed by the Victorian industrialists (Bradley, 1999: 69).

Economy of scale in the industrial era was the result of migration to the cities, abuse of female and child labour and workdays that were sometimes 20 hours long. These negative effects of accelerated economic growth, as a result of the Industrial Revolution, were described by Dickens in his 1854 novel *Hard Times* (Bradley, 1999: 70). The alienation from nature resulted in the alienation from the basic cell of every society—the family—as well as individuals from each other. Needs satisfaction was reduced to the minimum.

Industrialisation brought a collective spirit exclusively to industrialists in the field of profit making. The result beat the motive. National and global principles accelerated negative polarity between motives and consequences, which dialectically led to a new stage of socio-economic development.

According to Karl Marx:

> The worker becomes poorer if he/she produces more wealth, and if their production gets more power and a wider range. The worker becomes a cheaper commodity if he/she produces more commodities. The devaluation of the man of the world grows with the increase of the value of the world. Work does not only produce goods, but also itself and the worker as a commodity. (Quoted in Stojanović, 1987: 21)

7.3.3 *III Economy of information and knowledge*

The surplus value created in the economy and society as a result of industrialisation exceeds the boundaries of national states. The transport revolution and later the technological-informational revolution have marginalised the restrictive determinants such as time and space. As a result, economic relations have been internationalised and large corporations created. State intervention in this period is very powerful, since it is a necessary link in the process of internationalisation.

In the processes of internationalisation, the international institutions that paved the way to market liberalisation played a significant role, which led to a rapid expansion of capital and even alienation within companies (with the head office placed in one country and production in the other). Another consequence was a new alienation in distribution, so that on one side there were capital owners and on the other capital managers. The new phase of alienation affected not only the relation between workers and the results of their work but also the capitalists and their ability to manage their capital. Gigantism has its price. But despite the presence of adverse effects, the lower costs of transport, the ability to use low-cost inputs and information technology, as integration links, have all enabled the expansion and strengthening of the power of transnational companies. Capitalists create a new environment, the 'artificial nature' called the global market. In such an environment, artificially created resources become the most expensive, in order to eliminate risk and uncertainties that have become the axioms of the modern market. This artificial resource is known as information, and the economy based on it is known as the economy of information. Lack or unavailability of information is the result of insufficient knowledge and means the loss of positions in the market.

Thus, we come to the creation of a knowledge-based economy, which, in a dynamic environment, aims at the adoption of creative decisions.

> A fast and creative decision manages knowledge and encourages its conservation, engaging the creativity of other employees in the company, and at the same time responding to the changes in the environment.

Individualism in knowledge is more obvious at local levels. With the spread of knowledge, i.e. knowledge globalization, it becomes a more systematic category. In the category of the required capital in a multinational company, knowledge capital represents a new instrument in managing and creating the market position of the company. Modern conceptual study of knowledge differentiates between human and intellectual potential. Human potential implies a degree of employees' qualification; more precisely it represents explicit knowledge, while the intellectual capital represents innovation, creativity and ideas, i.e. tacit knowledge. Tacit knowledge is what the companies are fighting each other for today. Innovation and ideas create the competitive advantage in the market. Investing in knowledge means managing knowledge. Investing in knowledge capital means the investment in prosperity. No kind of capital has such significance as intellectual capital. The struggle among multinational companies is based on the struggle for intellectual supremacy. Establishing the knowledge market means investing in permanent capital. (Dragičević, 2005: 96).

7.3.4 *IV Brain economy*

Previous evolutionary-dialectical socio-economic processes have contributed to the creation of the new forces of production factors, such as information and knowledge, so that, with the previous marginalisation of space and time that was made possible owing to the transport and IT revolutions, local and national economic activities have become global. In that sense, information and knowledge become the variables with the global prefix, while the market also becomes globally competitive. By means of the synergetic operation of all listed variables, we come to the innovation that, with the accumulation of capital from the second phase of socio-economic relations (in our division), creates a new economy, the brain economy. Unlike the previous phase, where information and knowledge were generated, in this phase of the economy, information and knowledge are created by means of innovation; i.e., the market is created, bringing a new, sophisticated consumer and consequently a new consumer culture—*Meo voto* (according to my wish).

Global flows and the rapid dissemination of information and knowledge have led to major changes in the sphere of technology and information, which, in turn, have led to changes in the forms of the economy. Thus, from the economy based on the quantum of production, the economy of information and knowledge has been created, and today the brain economy, therefore creating a new framework of consumer culture.

Production is becoming increasingly specialised and sophisticated, with the new markets responding proactively to consumer demands. Being competitive and meeting the demands of increasingly demanding consumers are becoming synonymous of success. Modern consumers are becoming more passive participants in the market, compared with their more active participation in the past. Virtual business, which is a precondition of today's consumer culture, is becoming a new dimension of business and economic relations globally. In that regard, contemporary sociologists have noticed that today's young generations (born in the 1980s and 1890s) are marketing generations, for whom 'branded' products are of the greatest importance (Dragičević Radičević, Cvetković Bogavac and Gavrilović, 2012).

Modern consumer culture has materialised in the form of media, by means of the instruments of a much broader domain than the elements of the marketing mix once were. These new instruments can be recognised in the form of social responsibilities, advertising messages by emails, web pages, contests, virtual shopping, quizzes, various reality programmes, social networks and the like. It is interesting that today there are more marketing constraints than before (such as the ban on advertising alcohol and cigarettes), but the new forms of consumer culture propagation sometimes represent and allow media violence. The information revolution, with Internet and media technology being its central points, represents the basis of a new form of economy and the formation of today's consumer culture.

The consumer is not the one who makes the choice; rather, it is imposed on him. This makes it easier for companies to make profit. Profit is not realised through an individual. The basic rule of behaviour of economic subjects is to act rationally and to make rational decisions, whether as a consumer or a manufacturer. Adequate and relevant information and quality knowledge have the aim of minimising uncertainty and risk and enabling reliable planning for the future, better-quality decisions and expanded horizons of market choices. People constantly have some new requirements, which are the signal to producers to meet the market needs. The new consumer culture has new consumer needs, goals and methods of fulfilling these requirements. The expectations of market participants can often be unrealistic. Corporate giants in the form of transnational companies create the consumer spirit, which spreads rapidly; they do not leave the selling of their products and services to chance. Those giant companies could be described as the architects of contemporary consumer culture, the architects who, for short period of time, endeavour to meet consumer needs in a way that is convenient for them (Internet sales, armchair buying) and ensures their satisfaction.

In order to minimise the ambivalence of the process of globalisation, it is necessary to create a brain economy. All the subjects with creative thinking can contribute to increasing the degree of rationality in the decision-making process of all the various economic entities in the market. This can lead to the rational spread and usage of appropriate information and knowledge, as they are also the goods on the market, with their offer, demand, price, utility, cost and the like. On the other hand, a brain economy is necessary to rebuild the collective spirit in the spheres that should not be materialised and that must remain collective, since they represent the general good, such as morality, ethics and the like.

7.4 Information and knowledge vs. individualism and collectivism

During long-term global trends, the market has suffered major changes in terms of new forms of competition, reallocation of capital, a new mode of production, the more complex structures of the market, the needs for innovation skills of the existing workforce, the need for new job profiles and the like. As already known, economics and technology are in direct correlation, and it is therefore evident that the information revolution has introduced information and knowledge into the theory and practice of economics as innovated factors of production. Thus, the traditional division of factors necessary for production (labour, land, capital) has undergone changes and is now enriched with new factors—information and knowledge. The quantity of information and knowledge has improved the qualitative foundation of economics, defining the new form of economy based on creative thinking and action. Starting from the economy based on the quantum of production that marked the industrial era, we have come to the economy of information and knowledge within the framework of the post-industrial and information technology era, which has resulted in the era of innovation and a new form of economy—*the brain economy*.

It would be wrong to say that information and knowledge did not exist as important factors during the entire evolutionary process of society and economics as a science. Possessing the right information and adequate knowledge has always been the main objective of both individuals and society. It is only that their forms of manifestation and participation in the economy and life of people have changed with the times.

If we consider the developmental stages of different forms of economy, we can see that there are also the developmental stages of the use of information and knowledge. As information and knowledge passed through the transition process, the new forms of economy were gradually built.

Transition is a process that inevitably accompanies the process of globalisation. One of the main transitions is the economic one, which essentially implies a clearly defined strategy towards creating a competitive position in the international market. The restructuring of national economies is one of the preconditions for the creation of institutional conditions in the process of integration into global trends. In order for the economy to transition, it is necessary to define the existing environment, determine the key factors, define economic and political consequences and create adequate infrastructure (Dragičević Radičević, 2016).

The transition process of information and knowledge as essential factors of production contains five stages of development and change from the current to the desired condition of quality, definition and usage:

I   Individual use of information and knowledge in taming nature and customising natural scarce resources to satisfy the needs of the individual. As the biological, cultural and material evolution of people has permanently created new needs and as the number of people has been constantly increasing, the need for the right information and available knowledge has also been growing.

II Individual and collective use of information and knowledge in an adequate combination of factors of production (labour, land and capital). Due to the insufficiency of production factors and their alternative use, there is a problem of production structure and production methods and problems of the target group for which something is produced. Accordingly, this phase of the development of quality, definition and use of information and knowledge has certainly increased the level of importance of information and knowledge.

III The use based on collectivism and careful experimentation of individuals with the previously unexplored 'artificial nature'. This phase has given particular importance to adequate, timely and expensive information.

IV The phase that gives importance to the quality of knowledge and man's playing with 'artificial intelligence', which would certainly be impossible without the previously acquired high-quality knowledge. In today's business, a competitive position among individuals, companies and countries is acquired based on knowledge, as it increasingly becomes the foundation of and prerequisite for the creation of wealth, a perpetual economic motive and a subject of study. Knowledge has become available to everyone, anywhere and anytime. It is there, in an indefinite computer space, primarily due to the rapid development of the Internet. The revolution of knowledge is expanding at enormous speed. It cannot be isolated or enclosed or limited because information infrastructure enables it to expand to the world almost instantly. In that way, education has turned into a very effective 'technology' that follows contemporary changes. It is exactly people with their knowledge who make organisations different. Knowledge has become a priority and a strategic resource of companies.

V Since knowledge requires new ways of organising, business activities must be carried out in a different way from the traditional, which requires flexible and innovative forms of organization and will create conditions for a constant flow of creative achievements. More precisely, the last phase of economic development implies overcoming the traditional ways of thinking, behaving and organising of people who possess a higher degree of creative thinking. This phase is characterised by the play among people—i.e., the play with their existing knowledge and information. Creative thinking of every individual becomes important—but only through the group or society. The wealth of the society is no longer dependent on the ability to collect and convert raw materials but on the ability to accumulate the intellect of individuals and organisations and implement it in a creative and unique way.

Information and knowledge are significantly different from traditional productive forces, primarily because of their inexhaustible and unlimited character, as well as the inability to accurately measure the costs of creating 'information products'. The information revolution enables the separation of production from natural resources, capital from production, manufacturer from the production process, people's existence from physical labour and so on. Knowledge, information, skills, innovation and revolutionary ideas push the limits of social and economic growth and development, becoming thus the key wealth and productive forces.

7.5 Conclusion

At the beginning of the acceleration of the negative effects of modern global trends in the 20th century, Fromm underlined that the revolutionary changes that are necessary to humanize technical society, for the salvation of physical destruction, dehumanization and insanity, must be extended to all areas of life—economy, social life, politics and culture. In addition they must take place at the same time, because the partial change will not lead to a change of the whole system, and the symptoms of the disease break out in some other form. These changes are as follows:

a Changing production and consumption habits in such a way that economic activity focuses only on the development and growth of man, rather than on the current alienated system, transforms him, so that he is able to serve the goals of maximum production and maximum technical efficiency;

b Transformation of man as a citizen and participant in social life, in a sense that he is no longer allowed to be a passive, bureaucratic manipulated object, but makes him active, responsible and critical. In practice, this boils down to the revival of our management methods in a way that the political bureaucracy becomes subjected to effective control by the citizens, and involving in the process of decision-making in the private companies all those who work or who use their goods and services;

c The cultural revolution, which is trying to change the spirit of alienation and passivity characteristic for technical society, with the aim of creating a new man, whose life goals will not be the estate or spending wear. (1980: 155-156)

Historically, it is evident that through the evolutionary development of globalisation and socio-economic forms based on the use, quality and method of defining information and knowledge, there has been a diversity of motivation of the individual who has introduced individualism, even in the common good. The negative effects became more prevalent, which resulted in the dialectical process of 'negation of the negation' and led to a new chapter in socio-economic relations. Thus, new forms of the economy have emerged, such as the knowledge economy, which create more humane and qualitatively stronger individuals as part of the general social good.

Therefore, the pre-set assumption has been confirmed that due to the great social changes caused by dialectical developments and globalisation in society, complex changes of the socio-economic forms took place, which basically observe the individual and/or society as a whole. The dialectical process in the form of the Hegelian 'negation of the negation' has been confirmed, stating that with all the negative effects that today's socio-economic forms have, the new and higher form of economy, the so-called brain economy, is created.

References

Bošnjak, B. (1960). *Antologija filozofskih tekstova*, Zagreb: Školska knjiga.

Bradley, P. (1999). Victorian lessons: Education and utilitarianism in Bentham, Mill, and Dickens. *Concord Review*, *10*(2), 69-86.

Dragičević, T. (2005). *The impact of modern methods of business decision-making on the effectiveness and efficiency of enterprise in globalization conditions.* Belgrade: Megatrend University.

Dragičević Radičević, T. (2007). *International business.* Belgrade: Megatrend University.

Dragičević Radičević, T. (2014). Holistic economy vs. political individualism. In *Proceedings of the 6th International Scientific Conference* (pp. 45-55). Belgrade: Higher Educational Institution for Applied Studies of Entrepreneurship. ISBN 978-86-86707-604; COBBIS.SR-ID 207731468.

Dragičević Radičević, T. (2016). The impact of globalization on creating the maze of modern capital. In *Proceedings of the 8th International Scientific Conference* (pp. 180-188).

Dragičević Radičević, T., Cvetković Bogavac, N., and Gavrilović, M. (2012). *Economic and technological progress of civilization—From games with the nature to the game with the people.* Presentation at the International Conference on Social and Technological Development, 28-29 September 2012, Banja Luka, Republic of Srpska, University of Business Engineering and Management–Banja Luka.

Drucker. P. (1994, November). The age of social transformation. *Atlantic Monthly*, pp. 53-80.

Engels, F. (1945). *Anti-Dühring*. Zagreb: Naprijed.

Filipović, V. (1956). *Filozofija renesanse i odabrani tekstovi filozofa*, Zagreb: Matica Hrvatska.

Fischer, S. (2003). *Globalization and its challenges.* Peterson Institute for International Economics, https://piie.com/fischer/pdf/fischer011903.pdf.

Fromm, E. (1980). *Revolucija nade.* Beograd: Grafos.

Fromm, E. (1983). *Bekstvo od slobode.* Belgrade: Nolit.

Gavrilović, M. (2009). *The impact of globalization on the market reforms of the Serbian economy.* Master's thesis, Faculty of Business Studies, Megatrend University, Belgrade.

Jakšić, M. (2003). Globalization and the withdrawal of the state. In M. Knežević (Ed.), *Time of globalization* (p. 98). Belgrade: Dom kulture 'Studentski grad'.

More, T. (1952). *Utopija.* Beograd: Kultura.

Neef, M. M. (1991). *Human scale development: Conception, application and further reflections.* New York: Apex Press.

Nestorović, M. (2016). Different dimensions of the globalization process. In *Proceedings of the International Scientific Conference ERAZ 2016* (pp. 289-295). Belgrade: Association of Economists and Managers of the Balkans. ISBN 978-86-80194-03-5 (AEMB); COBISS.SR-ID 225322508.

Stojanović, I. (1987). *Neomarksizam i ekonomija.* Beograd: Ekonomika.

Stojanović, I. (2002). *Ekonomija.* Beograd: Megatrend Univerzitet.

Index